Mansfield Merriman

Elements of the method of least squares

Mansfield Merriman

Elements of the method of least squares

ISBN/EAN: 9783742891600

Manufactured in Europe, USA, Canada, Australia, Japa

Cover: Foto ©berggeist007 / pixelio.de

Manufactured and distributed by brebook publishing software (www.brebook.com)

Mansfield Merriman

Elements of the method of least squares

ELEMENTS

OF THE

METHOD OF LEAST SQUARES.

BY

MANSFIELD MERRIMAN, Ph.D.,

INSTRUCTOR IN CIVIL ENGINEERING IN THE SHEFFIELD SCIENTIFIC
SCHOOL OF YALE COLLEGE.

London:

MACMILLAN AND CO.

1877

[All Rights reserved.]

PREFACE.

In writing the following pages I have had two objects in view: first, to present the fundamental principles and processes of the Method of Least Squares in so plain a manner, and to illustrate their application by such simple and practical examples, as to render it accessible to Civil Engineers who have not had the benefit of extended mathematical training; and secondly, to give an elementary exposition of the theory which would be adapted to the needs of a large and constantly increasing class of students.

Hence the work falls into two parts, the first practical and the second theoretical, but each illustrating and supplementing the other. The numbering of the articles renders reference from one to the other easy; and the more thorough acquaintance the engineer makes with the second part the better will he adjust his observations, while it is only after much exercise with practical problems that the student can become thoroughly grounded in the theory.

Should the book, then, be taken up by students unfamiliar with the subject, let me suggest to them, that even if their aim be only to acquire a knowledge of its theory, the shortest and best way to do it is to become first familiar with the practical applications of Part I.; this attained, the rest follows naturally and easily.

As I have not written for mathematical experts, they will doubtless find considerable in the book at which to grumble. The idea of mean error does not appear. The term "equations of condition" has been, in accordance with the sensible German practice, divided into "observation equations" and "conditional equations" (*Beobachtungsgleichungen* and *Bedingungsgleichungen*), and each is used in its proper place. GAUSS' development of the law of probability of error has been followed as the best adapted to an elementary presentation, and if this be objected to as defective, I claim at least the credit of knowing and of pointing out (Art. 66) just what and where those defects are.

In preparing these pages I have consulted and freely used all the works upon the subject within my reach. The list of Literature and the historical notice at the end of the book will be of interest and value to all.

If this little elementary work should meet with a favourable reception from the scientific public, it may be followed by another containing extended applications of the method to higher geodetic surveying, and to numerous other problems arising in physical science, which have here been necessarily left unnoticed.

MANSFIELD MERRIMAN.

SHEFFIELD SCIENTIFIC SCHOOL,
  NEW HAVEN, CONN., U.S.A., *Feb.* 5, 1876.

CONTENTS.

PART I.

THE ADJUSTMENT AND COMPARISON OF ENGINEERING OBSERVATIONS 1

CHAPTER I.

INTRODUCTION 1

ART.
2—4.	Errors of Observations	2
5—9.	Principles of Probability	3
10—13.	The Probability Curve	8
13.	Table of Probability of Errors	12
14, 15.	The Method of Least Squares	14
16, 17.	The Comparison of Observations . . .	17
18—21.	Kinds of Observations	22

CHAPTER II.

DIRECT OBSERVATIONS UPON A SINGLE QUANTITY . 24

23.	The Arithmetical Mean	25
24—26.	Probable Error of a Single Observation and of the Arithmetical Mean	26
27—29.	Weights of Observations. The General Mean . .	30
30.	Probable Error of the General Mean	36
31.	Recapitulation	39

CHAPTER III.

INDEPENDENT OBSERVATIONS UPON SEVERAL QUANTITIES 41

33, 34.	Solution of Observation Equations	42
35.	Adjustment of Independent Observations of Equal Weight .	46
36, 37.	Observations of Unequal Weight	53
38—40.	Probable Errors and Weights	58
41, 42.	Other Applications	63

CHAPTER IV.

CONDITIONED OBSERVATIONS . . . 68

ART.
44, 45. Observations of Equal Weight 69
46. Adjustment of the Angles of a Quadrilateral . . . 76
47, 48. Observations of Unequal Weight 86
49. Probable Errors 94

CHAPTER V.

THE DISCUSSION OF PHYSICAL OBSERVATIONS . 100

51—54. The Deduction of Empirical Formulæ 100
55, 56. Probability of Errors. With Table 110
57. The Rejection of Doubtful Observations 116
58. Concluding Remarks 117

PART II.

THE THEORY OF LEAST SQUARES AND PROBABLE ERRORS 119

CHAPTER VI.

DEDUCTION OF THE FUNDAMENTAL PRINCIPLES . 119

5—9. Probability 120
10—13. Law of the Probability of Error 124
14, 15. The Principle of Least Squares 136
16, 17. The Measure of Precision and the Probable Error . . 138

CHAPTER VII.

DEVELOPMENT OF PRACTICAL METHODS AND FORMULÆ 141

22—31. Direct Observations upon one Quantity 141
32—42. Independent Observations upon several Quantities . . 153
43—49. Conditioned Observations 172
50—56. The Discussion of Observations 178

ART.		PAGE
	APPENDIX	183
59.	Observations involving non-Linear Equations . . .	183
60, 61.	Gauss' Method of solving Normal Equations	185
62.	Other Formulæ for Probable Errors	188
63.	The Mean Error	189
64.	List of Literature	190
65.	On the History of the Method of Least Squares . . .	194
66.	Remarks on the Theory of Least Squares	196
	ALPHABETICAL INDEX	199

CORRECTIONS.

Page 17, line 9 from foot, *for* (29) *read* (28).
" 53, " 7 from top, *for* 6·9 *read* 6·1.
" 56, " 2 from foot, *for* 36·3 *read* 36·64.
" 56, " 1 from foot, *for* 11·7 *read* 11·1.
" 56, " after last line, *insert* $AOC = 88° 33' 41''·7$ weight 2;
" 57, " 12 from top, *for* $-0·7$ *read* $-0·36$.
" 57, " 13 from top, *for* $-0·3$ *read* $-0·9$.
" 57, after line 13, *insert* $z' + y' = -0·3$ with weight 2.
" 76, lines 15—18 from top, *for* s', t', u', y', z', *read* s, t, u, y, z.
" 80, line 6 from top, *for* 8·39 *read* 0·39.
" 81, " 1 from foot, *for* sin WYZ sin WZY *read* sin WYX sin WZY.
" 81, " 19 from foot, *for* WYZ *read* WYX.
" 89, equations (110), *for* y_1, y_2, y_3, *read* γ_1, γ_2, γ_3.
" 92, line 15 from top, *for* Which *read* which, and *dele* ?.
" 94, lines 8—15 from top, *for* ρ *read* p, and *for* ν *read* v.
" 96, line 16 from top, *after* conditional *insert* equations.
" 101, " 3 from foot, *for* many observations *read* many equations.
" 112, *for* ·03228, ·62679, ·97573 *read* ·04303, ·62671, ·97657.
" 114, line 3 from foot, *for* 22 *read* 605.
" 115, " 14 from top, *for* 17 *read* 25.
" 115, " 11 from foot, *for* $+0''·8$ *read* $\pm 0''·8$.
" 117, " 10 from top, *for* works Nos. 28 and 34 *read* work No. 38.
" 135, " 2 from foot, *for* McLaurin's *read* Maclaurin's.
" 143, equation (32), *for* $P - y_1 y_2 \ldots$ *read* $P = y_1 y_2 \ldots$.
" 144, lines 4—5 from foot, *for* c *read* c'.
" 145, line 1 from top, *for* c *read* c'.
" 172—6, *for* ρ *read* p.

PART I.

THE ADJUSTMENT AND COMPARISON OF ENGINEERING OBSERVATIONS.

CHAPTER I.

INTRODUCTION.

ART. 1. When a quantity is observed with a view to determining its magnitude a *number* is obtained as the result of the operation. This number expresses how many units and parts of a unit are, according to the measurement, contained in the observed quantity, and is hence a measure of its magnitude. The word *observation* will be used in this book to express such numerical measures as well as the operation by which they are deduced.

Every engineer is cognizant of the fact, that, when several observations or sets of observations are made to determine the magnitude of a quantity (for example, the length of a line) the results do not agree. Since the quantity can have only one value all of these results cannot be correct, and each one of them can be regarded only as an approximation to the truth. The absolutely true value of the quantity in question, we can never obtain or at least be never sure that we have obtained, and instead of it we must accept and use a value, derived from the combination of our observations, which may perhaps not exactly agree with any one of them, but which however shall be the *most probable* value (Art. 9).

Part I. of this work will be devoted to the presentation and illustration of the methods now in use among scientific men for the adjustment and comparison of observations, and the determination of the most probable values of observed quantities, with sufficient mathematical reasoning to exemplify the main principles of these methods in a clear light to the general reader. Part II. will contain a full development and discussion of the theories upon which they are founded. The first is designed more particularly for the use of those who have little time to devote to the niceties of mathematical theory, but who are desirous of learning the fundamental principles and practical applications of the science; the second is intended for the use of students taking up the subject from a theoretical point of view. Each part, however, will be to some extent dependent upon the other, and hence for convenience of reference the corresponding articles and formulæ will be marked by corresponding numbers.

Errors of Observations.

2. *Constant Errors* are those which always under the same circumstances have the same value, and which therefore strictly speaking are not errors, but the results of law. As such we may mention: *theoretical* errors like the effects of refraction in increasing the size of a vertical angle, or the effects of temperature upon the length of rods used in measuring a base line, which effects when their causes are understood may be computed beforehand, and are hence no longer to be classed as errors; *instrumental* like those arising from an incorrect graduation of the limb of a theodolite, or an imperfect adjustment of the line of collimation in a telescope, which may be also removed by calculation or by a proper mode of using the instrument; and *personal* errors which arise in very delicate observations and are due to the habits of the observer, who may for instance in reading a vernier always give the number of records too large or too small by a constant quantity, and which may be corrected by the application of a "personal equation."

Such errors being capable of elimination need no further consideration in the discussion of our subject.

3. *Mistakes* are a class of errors committed by inexperienced and occasionally even by the most skilled observers, arising from mental confusion. As such are; mistakes in

measuring an angle arising from sighting at the wrong signal; mistakes in reading an angle by noting 54° instead of 46°, etc. This class of errors sometimes admits of correction by comparison with other sets of observations: it will receive no further notice in this book.

4. *Accidental Errors* are those that still remain after all constant errors and mistakes have been carefully investigated and eliminated from the numerical results. Such, for example, are the errors in levelling arising from sudden expansions and contractions of the instrument or from the effects of the wind, or those often observed in sighting across a river arising from the anomalous and changing refraction of the atmosphere. More than all, however, such errors arise from the imperfections of our touch and sight, which render it impossible for us to handle our instruments delicately, or estimate accurately small divisions of their graduation. These are the errors which appear in our numerical results, however carefully the measurements be made, and which form the subject of the following pages. Although at first sight it would seem that such irregular errors could not come within the province of mathematical investigation, it will be seen in the sequel that they are governed by a wonderful and very precise law, viz. the law of probability (Art. 11).

Principles of Probability.

5. We must therefore, as preliminary to our subject, state and exemplify the mathematical definition of probability and of the words "most probable" which we have used in Art. 1.

If a coin be tossed up into the air, it is said in common language that the chances are equal that it will turn up head or tail, or that the occurrence of head or tail is equally probable. So if a die marked in the usual manner be thrown, it is said that the odds are one to five in favour of throwing the ace, or five to one against throwing the ace. This is the statement in popular language of the idea of probability; its mathematical expression is but slightly different. In throwing the coin we recognise that there are two possible cases, *either* head *or* tail may turn up, and one is as likely to occur as the other; and hence we express the probability of throwing a head by the fraction $\frac{1}{2}$, and the probability of throwing a tail also by $\frac{1}{2}$. So with the die there are six equally probable cases, one of which may be the ace, and hence the probability of throwing the ace in one trial is $\frac{1}{6}$, and the probability of not throwing it is $\frac{5}{6}$. *The probability of the occurrence of an event is,* then, *always a fraction whose denominator denotes the whole number of possible cases* (each supposed equally probable) *and whose numerator denotes the number of cases favourable to its occurrence.*

Thus if there be in a bag 30 red and 20 white balls, and a ball is drawn at random, it will be either red or white; both events are *equally possible* but *not equally probable;* for the probability of drawing a red ball is $\frac{30}{50}$ and that of drawing a white $\frac{20}{50}$, or $\frac{3}{5}$ and $\frac{2}{5}$ respectively.

The mathematical expression of the probability of the occurrence of an event is, then, a numerical measure of our degree of confidence that it will occur. As a fraction may have any value from 0 to 1, so our confidence may range from *impossibility* to *certainty;* a small fraction like $\frac{1}{1024}$ denotes a very small probability, and a large one like $\frac{1023}{1024}$ denotes a very large probability or almost a certainty.

6. If there be 18 red balls in a bag, and one is drawn out, the probability that it is red is $\frac{18}{18}=1$, that is, it is *certain* that such a one will be drawn. Therefore: *Unity is the mathematical symbol for certainty.* Hence if the probability of the happening of an event be known, the probability that it will not happen is unity minus the first probability.

INTRODUCTION. 5

Thus the probability of throwing an ace in one trial is $\frac{1}{6}$, the probability of *not* throwing it is $1-\frac{1}{6}=\frac{5}{6}$, as shown otherwise in the preceding article.

7. If there be in a bag 20 red, 16 white, and 14 black balls, and one is to be drawn out, the probability that it will be red is $\frac{20}{50}$, that it will be white $\frac{16}{50}$, and that it will be black $\frac{14}{50}$. If however we ask the probability of drawing *either* a red or white ball, we have 36 favorable cases out of the 50 total cases, and the answer is $\frac{36}{50}$ or $\frac{20}{50}+\frac{16}{50}$. Hence, *if an event may happen in different independent ways, the probability of its happening is the sum of the separate probabilities.* Thus in tossing a coin, the probability of throwing *either* head or tail is $\frac{1}{2}+\frac{1}{2}=1$, that is, *one* is certain to be thrown.

8. Let there be two bags, one of which contains 7 black and 9 white balls, and the other 4 black and 11 white balls. The probability of drawing a black ball from the first bag is $\frac{7}{16}$, that of drawing one from the second $\frac{4}{15}$. What now is the probability if I draw from both bags at the same time that both balls drawn will be black? Since each ball in the first bag may form a pair with each one in the second, there are 16×15 possible ways of drawing two balls; and since each of the 7 black balls may be associated with each of the 4 black balls to form a pair, there are 7×4 cases favorable to drawing two black balls. The required probability is hence $\frac{7 \times 4}{16 \times 15}$ and since this is equal to $\frac{7}{16} \times \frac{4}{15}$ we have the principle that *the probability of the happening of several independent events is equal to the product of their respective probabilities.*

INTRODUCTION.

Thus if three dice be thrown, the probability that all will be aces is $\frac{1}{6} \times \frac{1}{6} \times \frac{1}{6} = \frac{1}{216}$, a small fraction.

9. Suppose two coins to be thrown up at the same time, they may both turn up heads or both tails, or one may be a head and the other a tail. We wish to determine the respective probabilities. Let us call the two coins A and B, then the cases which may happen are

1. A head and B head,
2. A head and B tail,
3. A tail and B head,
4. A tail and B tail;

and each of these cases being equally likely to occur, has for its probability $\frac{1}{4}$. Hence the probability that both will be heads is $\frac{1}{4}$, that one will be head the other tail $\frac{1}{4} + \frac{1}{4} = \frac{1}{2}$, and that both will be tails $\frac{1}{4}$. The sum of the probabilities $\frac{1}{4} + \frac{1}{2} + \frac{1}{4}$ is unity, as should be the case (Art. 6), since one of the events is certain to occur.

In like manner if there be ten coins thrown up at the same time there may occur the following groups, having the respective probabilities as annexed. (For an easy method of computing these probabilities for 10 or any number of coins, see Part II., Art. 9.)

All 10 coins may be heads, ... probability $\frac{1}{1024}$;

9 may be heads and 1 tail, ... $\frac{10}{1024}$;

8 .... 2 .... $\frac{45}{1024}$;

7 .... 3 .... $\frac{120}{1024}$;

6 may be heads and 4 tails, probability $\frac{210}{1024}$;

5 .... 5 .... $\frac{252}{1024}$;

4 .... 6 .... $\frac{210}{1024}$;

3 .... 7 .... $\frac{120}{1024}$;

2 .... 8 .... $\frac{45}{1024}$;

1 .... 9 .... $\frac{10}{1024}$;

All may be tails .... $\frac{1}{1024}$.

The sum of all these probabilities is unity, since one of the groups is certain to occur. In speaking of this case in popular language, it is said that it is very improbable that all the coins will be heads, in mathematical language we say that the probability is small, viz. $\frac{1}{1024}$; so both common and mathematical reasoning recognise that the group of 5 heads and 5 tails is the *most probable*.

In the theory of probabilities, therefore, we say, *an event is the most probable when the* a priori (*or theoretical*) *probability of its happening is greatest among all the probabilities of all the possible events*, that is, *when that probability is a maximum*. Thus if two persons of exactly equal skill should play 51 games of cards, it is possible but not probable that one of them will win all the games; the *most probable* case is as every one will recognise, that one will win 25 and the other 26 games.

We advise the reader, before proceeding further, to apply the above principles of probability to the solution of the following problems.

Problems. 1. A lottery has 250 tickets, 25 of which are prizes, and the remainder blanks. What is the probability of not drawing a prize?

2. A bag contains 100 balls, 50 are red, 12 are white, and the rest are black. What is the probability of drawing *either* a white or a black ball in one trial?

3. The probability of throwing an ace with a single die in two trials is $\frac{11}{36}$. What is the probability of not throwing it?

4. The probability that A can solve a certain problem is $\frac{2}{3}$, the probability that B can solve it is $\frac{1}{4}$. What is the probability that it will be solved if both try? *Ans.* $\frac{3}{4}$.

5. A bag contains 3 red, 4 white, and 5 black balls. Required the probability of drawing 2 red balls, *the ball first drawn not being replaced before the second trial.* *Ans.* $\frac{1}{22}$.

6. Six coins are tossed up. What is the probability that 4 of them will be heads and 2 tails? *Ans.* $\frac{15}{64}$.

7. A coin is tossed up four times in succession. What is the probability that all will be heads?

The Probability Curve.

10. We may now proceed to consider the law of the probability of errors of observation.

If a person accustomed to the use of the rifle, shoot 500 times at a target, all the balls will not hit the central bull's-eye, and some perhaps will not even hit the target. The deviation of each bullet from the centre of the target is an *error*, and furthermore an accidental error (Art. 4) produced by changes in the wind, imperfections in the aim of the marksman, etc.; for all constant errors, such as the effect of gravitation, are assumed to be eliminated in the sighting of the rifle. An examination of the bullet marks on the target shows us, however, that these errors are arranged around the central point in a very regular and symmetrical manner. First we observe that small errors (that is, deviations from the centre) are more frequent than large ones; secondly that they are arranged symmetrically around the centre so that at equal distances, above, below, and in every direction from that centre, the number of marks in a square inch is the same; and thirdly we recognise the fact that very large errors, for instance, a deviation from the centre of half a mile, do not occur. Further, we observe that the greater the skill of the marksman, the nearer are the marks to the centre point.

Again, suppose an engineer to measure an angle a hundred times, each time with equal care. The readings will disagree, the differences between the true value of the angle and each of his results will be an error; and we recognise that these errors, like those of the marksman, will be subject to three laws; viz. 1st, small errors will be more frequent than large ones; 2nd, errors of excess and deficiency (that is, results greater and less than the true value) will be equally numerous, and 3rd, large errors like those of 2° do not occur. Further, the more skilled the engineer in the measurement of angles, the nearer will be his readings to the true value of the angle, and the smaller will be his errors.

These three axioms or fundamental laws form the foundation of the science of the adjustment of observations.

11. In any set of carefully made observations, then, the probability (Art. 5) of a small error is greater than that of a large one, the probability of an error in excess of the true value is the same as that of an equal error in deficiency, and the probability of a very large error is zero. Thus the probability of an error is a *function* of the error, so that if x represent any error, and y its probability, we may express the relation between these quantities by the equation

$$(1) \quad y = f(x),$$

which is read, y equals a function of x, that is, y is dependent upon x for its value.

If we take then y as an ordinate, and x as an abscissa, we may regard this as the equation of a curve, which must be of a form so as to agree with the three laws adopted above; viz., its maximum ordinate OA must correspond to the error 0; it must be symmetrical with respect to the axis of Y, since equal positive and negative errors are equally probable, as x increases numerically the value of y must rapidly decrease,

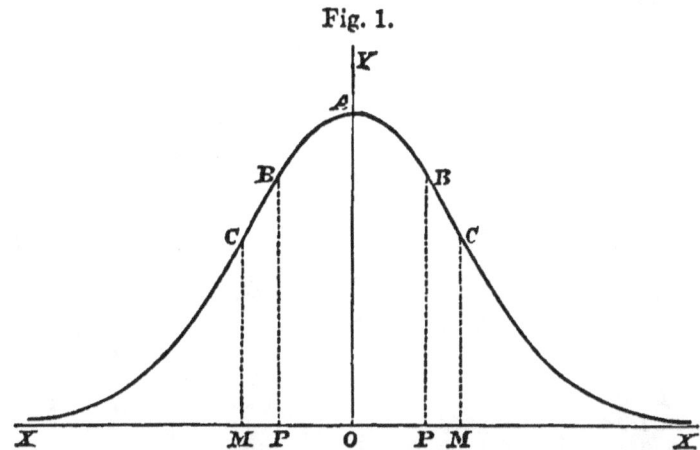

Fig. 1.

and when x becomes very large y must be zero. Fig. 1 represents such a curve, OP and OM being errors, and PB and MC their corresponding probabilities. Further, since different observations have different degrees of accuracy, each set of observations will have a distinct curve of its own, whose particular dimensions will depend upon the precision of the measurements, but whose general form must be that of Fig. 1.

In Part II. we have given GAUSS' method of determining the equation of this curve from the principles of the theory of probabilities (Arts. 5, 9), the fundamental axioms of Art. 10, and the rule of the arithmetical mean or average (Art. 23). Its equation is

$$y = \frac{c}{e^{h^2 x^2}},$$

or as it is usually written

$$(12) \quad y = ce^{-h^2 x^2},$$

in which e denotes the base of the Napierian system of logarithms or 2·71828, and c and h are constants which depend upon the accuracy of the observations, and which determine the dimensions of the curve for any particular case. Regarding for the present both c and h as unity, and attributing values to x, we may by logarithms find the corresponding values of y. The following table gives a few of these values, from which the reader may construct a curve to scale and observe its correspondence in form with that given in Fig. 1.

$$y = e^{-x^2} = \frac{1}{e^{x^2}}.$$

x	y	x	y
·0	1·0000	±1·8	0·0392
±0·2	0·9608	±2·0	0·0183
±0·4	0·8521	±2·2	0·0079
±0·6	0·6977	±2·4	0·0032
±0·8	0·5273	±2·6	0·0012
±1·0	0·3679	±2·8	0·0004
±1·2	0·2370	±3·0	0·0001
±1·4	0·1409		
±1·6	0·0773	±∞	0·0000

This is called the *Probability Curve*, and its equation is the mathematical expression of the law of the probability of error, upon which the whole science of the adjustment and comparison of observations, as here developed, is founded.

12. This curve has many curious properties, only one of which we can notice here.

If we measure an angle 500 times there will probably be 250 readings greater than the true value and 250 less. Subtracting each reading from the true value we have the errors, 250 being positive and 250 negative. Now if we lay off each of these errors as an abscissa x, and erect its corresponding ordinate y, we shall have a curve of the form of Fig. 1. Now the property of the curve to which we wish to call attention is this: If we represent the total area between the curve and the axis of X by 500, the area $OABP$ on the right of the axis Y denotes the number of positive errors less than OP, the area $PBCM$ denotes the number greater than OP and less than OM, and the area CMX denotes the number greater than OM: likewise the area $OABP$ on the left of the axis Y denotes the number of negative errors less than OP, and so on; also, if $OP = OP$, the area $PBABP$

exhibits the number of errors less numerically than OP. In Fig. 1 we have drawn the ordinates so that the areas $OABP$, $PBCM$, and MCX represent 125, 45, and 80 errors respectively.

Probability of Errors. (22) $P' = \dfrac{2}{\sqrt{\pi}} \int_0^{hx} e^{-h^2 x^2} d(hx)$.

hx	P'	hx	P'	hx	P'	hx	P'
0·00	0·00000	0·60	0·60386	1·20	0·91031	1·80	0·98909
0·02	·02256	0·62	·61941	1·22	·91553	1·82	·98994
0·04	·04511	0·64	·63458	1·24	·92050	1·84	·99073
0·06	·06762	0·66	·64938	1·26	·92523	1·86	·99147
0·08	·09008	0·68	·66378	1·28	·92973	1·88	·99216
0·10	0·11246	0·70	0·67780	1·30	0·93401	1·90	0·99279
0·12	·13476	0·72	·69143	1·32	·93806	1·92	·99338
0·14	·15695	0·74	·70468	1·34	·94191	1·94	·99392
0·16	·17901	0·76	·71754	1·36	·94556	1·96	·99443
0·18	·20093	0·78	·73001	1·38	·94902	1·98	·99489
0·20	0·22270	0·80	0·74210	1·40	0·95228	2·00	0·99532
0·22	·24429	0·82	·75381	1·42	·95537		
0·24	·26570	0·84	·76514	1·44	·95830		
0·26	·28690	0·86	·77610	1·46	·96105		
0·28	·30788	0·88	·78669	1·48	·96365		
0·30	0·32863	0·90	0·79691	1·50	0·96610	3·00	0·99998
0·32	·34912	0·92	·80677	1·52	·96841		
0·34	·36936	0·94	·81627	1·54	·97058		
0·36	·38933	0·96	·82542	1·56	·97263		
0·38	·40901	0·98	·83423	1·58	·97455		
0·40	0·42839	1·00	0·84270	1·60	0·97635		
0·42	·44747	1·02	·85084	1·62	·97804		
0·44	·46622	1·04	·85865	1·64	·97962		
0·46	·48465	1·06	·86614	1·66	·98110		
0·48	·50275	1·08	·87333	1·68	·98249		
0·50	0·52050	1·10	0·88020	1·70	0·98379		
0·52	·53790	1·12	·88679	1·72	·98500		
0·54	·55494	1·14	·89308	1·74	·98613		
0·56	·57161	1·16	·89910	1·76	·98719		
0·58	·58792	1·18	·90484	1·78	·98817	∞	1·00000

If, then, we express the total area of the curve by unity, the areas $PBABP$, $MCACM$, etc. corresponding to the successive values OP, OM, etc. will be fractions, proportional to the number of errors less than those values of x. The preceding table, which may be computed by the methods of Part II., is taken from the *Berliner Jahrbuch* for 1834, and gives the areas on both sides of the axis Y corresponding to successive numerical values of hx, x being the error, and h the measure of precision, referred to in Art. 11.

13. To use this table it is necessary to know the value of the constant h. Methods will be hereafter given (Arts. 24, 38) by which its value may be found for any given observations. Granting for the present that it may be determined, the following example will show the use of the table, and exemplify the accordance of theory and practice.

BESSEL in his *Fundamenta Astronomiae* discusses 470 observations made by BRADLEY upon the right ascensions of the stars Sirius and Altair, and determines the measure of precision to be $h = \dfrac{1}{0''{\cdot}5529}$. Now let it be required to find the number of errors less than $0''{\cdot}2$, $0''{\cdot}4$, $0''{\cdot}6$, etc. For the number less than $0''{\cdot}2$ we must take $x = 0''{\cdot}2$, and

$$hx = \frac{0''{\cdot}2}{0''{\cdot}5529} = 0{\cdot}3616;$$

and for the number less than $0''{\cdot}4$, $0''{\cdot}6$, etc., hx will be the successive multiples of $0{\cdot}3616$ by 2, 3, etc. We find then from the table,

for $x = 0''{\cdot}2$ with $hx = 0{\cdot}362$ the area $P' = 0{\cdot}39102$,

... $x = 0{\cdot}4$ ...... $hx = 0{\cdot}723$ .... $P' = 0{\cdot}69372$,

... $x = 0{\cdot}6$ ...... $hx = 1{\cdot}085$ .... $P' = 0{\cdot}87511$,

... $x = 0{\cdot}8$ ...... $hx = 1{\cdot}447$ .... $P' = 0{\cdot}95926$,

.. $x = 1{\cdot}0$ ...... $hx = 1{\cdot}808$ .... $P' = 0{\cdot}98946$,

... $x = \infty$ ...... $hx = \infty$ .... $P' = 1{\cdot}00000$.

Now from the property of the curve above noticed these values of P' are proportional to the number of errors less than the corresponding value of x. Hence multiplying each of these numbers by 470, the total number of errors, and comparing them with the actual number of errors as given by BESSEL, we have

theoretical no. of errors less than 0″·2 is 184, actual no. was 182,
.... 0·4 ... 326, ...318,
.... 0·6 ... 412, ...405,
.... 0·8 ... 451, ...445,
.... 1·0 ... 465, ...462,
.... over 1·0 ... 5, ... 8.

The agreement between theory and experience though very close is not exact, partly, perhaps, because what we have called the actual number of errors were computed not from the *true* values of the right ascensions of the stars but from their *most probable* values as deduced from the 470 observations.

The number of errors between any two given limits is of course found by simple subtraction; thus, in the above example there are 142 theoretical errors between 0″·2 and 0″·4, 86 between 0″·4 and 0″·6, 39 between 0″·6 and 0″·8, and 14 between 0″·8 and 1″·0.

Problems. 1. A marksman shoots 500 times at a target. If his skill is such that when the errors x are measured in feet h is unity, what are the number of bullet-marks between two circles described from the centre whose radii are 1 and 2 feet? *Ans.* 76.

2. An angle is measured 100 times, and the value of the measure of precision found to be $h = \frac{1}{5''}$. How many errors are greater than 2″ and less than 4″? How many greater than 8″?

This subject is further considered in Art. 55.

The Method of Least Squares.

14. Suppose now a number of observations be made to determine the values of certain quantities, for example, the lengths of lines. However carefully the measurements be made there will be disagreement in the results, and hence

INTRODUCTION. 15

we can never be sure that any adjustment that we may make will give us the absolutely true values of the lines. The most we can do is to determine approximate results, which shall be the *most probable* values (Art. 9), and moreover be rendered the most probable value by the existence of the observations themselves.

The preceding principles of the probability of error afford us a general rule for the determination of the most probable values of observed quantities. To deduce it in its simplest form, let us suppose that the observations are equally good, that is, that they are made with exactly the same precision; then the constants c and h (Art. 11) which measure that precision will be the same for all observations. Let us designate by s, t, u, etc. the quantities whose values are to be derived from the measurements, and by M_1, M_2, M_3, etc. the results of those measurements, which are made either directly (Art. 18) upon s, t, u, etc., or indirectly (Art. 19) upon other quantities related to them. If our measurements were perfect, M_1, M_2, etc. would be absolutely true values, and there would be no discordance in the results found for s, t, u, etc.; but being imperfect the results do not agree, and hence *errors* (Arts. 4, 10) exist. Let those errors, which are the differences between the measurements M_1, M_2, M_3, etc., and the corresponding *true* values be denoted by x_1, x_2, x_3, etc.

Now if x be *any* error and y its corresponding probability, the law of the probability of errors (Art. 11) gives us

(12) $\quad y = ce^{-h^2x^2}$.

Hence for the errors actually committed we have

Prob. of the error $x_1 \doteq y_1 = ce^{-h^2x_1^2}$,
.... $x_2 \doteq y_2 = ce^{-h^2x_2^2}$,
.... $x_3 \doteq y_3 = ce^{-h^2x_3^2}$, etc.

Now by Art. 8 the probability of committing all these errors is the product of these respective probabilities: hence it is

$y_1 y_2 y_3$ etc. $= ce^{-h^2x_1^2} \times ce^{-h^2x_2^2} \times ce^{-h^2x_3^2} \times$ etc.

Designating this product by P, the expression becomes

(23) $\quad P = c \times c \times c \times$ etc. $\times e^{-h^2(x_1^2+x_2^2+x_3^2+x_4^2+\text{etc.})}$,

in which c, e and h are constants. Now what the *true* values of s, t, u, etc. are, we cannot hope to find, and must hence be content with determining their *most probable* values; and the most probable system of values is that which has the greatest probability (Art. 9). Each of the errors, x_1, x_2, x_3, etc., is dependent upon (is a function of) the quantities s, t, u, and the most probable system of values for the latter corresponds to the most probable system of errors.

The most probable system, then, is that for which P is a *maximum* (Art. 9), and P in the above expression will be a maximum when the exponent of e is a *minimum*, that is, when

(24) $\quad x_1^2 + x_2^2 + x_3^2 + x_4^2 +$ etc. $=$ a minimum,

and as these terms are the squares of the differences or errors, we have the principle that, *the most probable values of quantities, which are the object of measurement, are those which render the sum of the squares of the errors a minimum.*

15. To illustrate the application of this principle, let us suppose that the measurements M_1, M_2, M_3, etc. are made directly upon the *same* quantity M whose *true* value is z. Then we have committed the errors

$$(z - M_1), (z - M_2), (z - M_3), \text{etc.}$$

Now by the preceding principle the most probable value for z is that which renders

(24) $\quad (z - M_1)^2 + (z - M_2)^2 + (z - M_3)^2 +$ etc.

a minimum. If n be the number of observations, the last term of this expression will be $(z - M_n)^2$. Applying the usual method for determining maxima and minima, we differentiate the expression thus:

$$2(z - M_1)dz + 2(z - M_2)dz + 2(z - M_3)dz + \ldots + 2(z - M_n)dz;$$

place it equal to zero and divide the equation by $2dz$, giving

$$(z - M_1) + (z - M_2) + (z - M_3) + \ldots + (z - M_n) = 0.$$

Solving this equation and denoting the resulting value of z by z_0, we have

$$z_0 = \frac{M_1 + M_2 + M_3 + \ldots + M_n}{n},$$

that is, the most probable value of z is the average of the n measurements (Art. 23)*.

From the principle of Art. 14 arises the name "Least Squares" to express the method now universally in use for the *adjustment* of observations.

The Comparison of Observations.

16. In Art. 10 we observed that the greater the skill of the marksman the nearer are the bullets to the centre of the target, and also that the more skilled the engineer, the less will be the deviations of his readings from the true value of the angle. In Art. 11 our law of the probability of error,

$$(12) \qquad y = ce^{-h^2x^2},$$

contained two constants, c and h, dependent upon the precision of the observations. The accuracy of different sets of observations may then be compared by comparing their measures of precision given by h, or by comparing other constants related to h. The one usually employed for this purpose is called the *probable error r,* which we define as *the error which has such a value that the number of errors greater than it is the same as the number less than it.* Thus in the probability curve, Fig. 1, the total area $XBABX$ denotes the total number of errors in a set of observations (Art. 12), and if the area $PBABP$ be one-half of that total area, one-half of the errors will be less than OP, and hence by our definition OP is the probable error. Now since r is of such a value that if $r = OP$, the area $PBABA$ is $\tfrac{1}{2}$ when the total area is unity, we have only to find from the table (Art. 13) the value of $hx = hr$ for which $P' = 0.5$. By interpolation between the values $hx = 0.46$ and $hx = 0.48$, we find that $P' = 0.5$ when $hx = 0.4769$. Hence we have

$$(29) \qquad hr = 0.4769, \text{ or } r = \frac{0.4769}{h},$$

* The above is not strictly a demonstration of the law of the average or arithmetical mean, for the equation of the probability curve (Art. 11) is deduced upon the assumption that in direct observations the average is the most probable true value; and hence the above is really reasoning in a circle. It serves, however, to illustrate to the reader that the principle of least squares is in agreement with the universally adopted method of taking the arithmetical mean of equally good observations. See Art. 66.

from which we see that the values of h and r are always reciprocally proportional.

To render more definite our conception of the measure of precision h and the probable error r, let us consider the case of two sets of observations made with different degrees of accuracy. Let the measure of precision of the first be h_1, and of the second h_2; then for the probability of errors in the first set we shall have a curve whose equation is

$$(19) \quad y = h_1 i \pi^{-\frac{1}{2}} e^{-h_1^2 x^2},$$

and for the second a curve given by

$$(19) \quad y = h_2 i \pi^{-\frac{1}{2}} e^{-h_2^2 x^2},$$

in which i is a very, very small constant, independent of h, and π is the number 3·1416. Now let us suppose that the second set is twice as accurate as the first, so that $h_1 = h$ and $h_2 = 2h$; then the equations will be

$$y = h i \pi^{-\frac{1}{2}} e^{-h^2 x^2} \text{ and } y = 2 h i \pi^{-\frac{1}{2}} e^{-4 h^2 x^2}.$$

If for purposes of comparison we consider h and i as unity and compute the values of y corresponding to successive

Fig. 2.

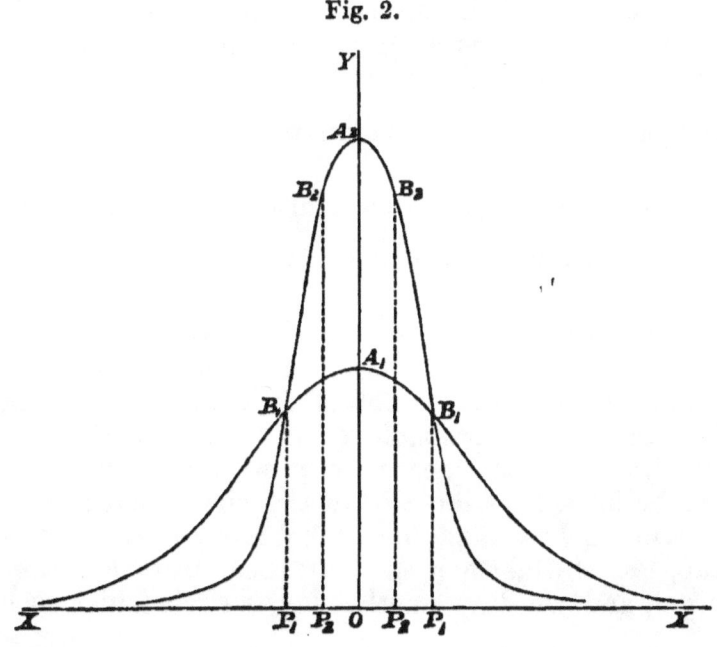

numerical values of x, we may obtain two tables similar to that of Art. 11, from which the two corresponding curves may be plotted. Fig. 2 shows these curves, $XB_1A_1B_1X$ being the one for the set of observations whose measure of precision is h_1 or h, and $XB_2A_2B_2X$ the one for the set whose measure of precision is h_2 or $2h$. These curves show at a glance the relative probabilities of corresponding errors in the two sets; thus the probability of the error 0 is twice as much in the second as in the first set, the probability of the error OP_1 is nearly the same in each, while the probability of an error twice as large as OP_1 is much smaller in the second than in the first set. Now if we draw the lines P_1B_1, P_2B_2, so that the areas $P_1B_1A_1B_1P_1$ and $P_2B_2A_2B_2P_2$ are respectively one-half of the total areas of their corresponding curves, the line OP_1 will be the probable error of an observation in the first set, and OP_2 the probable error of one in the second set. If we represent these by the letters r_1 and r_2, we must in each case have the constant relation

(28) $\qquad h_1 r_1 = 0{\cdot}4769, \quad h_2 r_2 = 0{\cdot}4769$;

and since h_2 is twice h_1, it follows that r_2 must be one-half of r_1, as is represented in Fig. 2.

Thus we see that the probable error is an error of such a magnitude that it is as likely as not that any assigned error will exceed it or fall short of it. Hence the probability that any error x taken at random will be less than r is $\frac{1}{2}$, and that it will be greater than r also $\frac{1}{2}$. It is then an even wager that any error x will be greater or less than the probable error r.

Referring now to the case of the 500 bullet-marks on the target, let us imagine a circle described from the centre which shall include exactly 250 of those marks. The radius of this circle is the probable error of the marksman. If he is to aim and shoot once more, it will be an even chance that the bullet will strike within the circle, that is, that his error x will be less than his probable error r. If another rifleman, less skilful than the first, shoot also 500 bullets at the target, and we draw another larger circle which includes

250 of them, the radius of this circle will be *his* probable error. If the circle of the first marksman be 4 inches and that of the second 12, we recognise from our constant relation between h and r, that the precision of the first is three times that of the second; and while we can afford to wager one to one that a shot of the first man will fall within 4 inches of the centre, we can only for the second man afford to make an even wager that it will fall within 12 inches of the centre.

Methods will be hereafter given (Art. 24) by which h, and consequently r, may be determined for any given observations. For the present we give the following numerical example, illustrating more fully the practical use of the probable error.

17. Let an angle be measured ten times equally carefully by a theodolite reading to 10″, and again be measured the same number of times with a transit reading only to 1′. Suppose the results to be the following:

Observation.	By Theodolite.			By Transit.	
	°	′	″	°	′
1.	24	13	40	24	13
2.		13	10		14
3.		13	30		15
4.		13	40		13
5.		14	0		12
6.		13	20		13
7.		13	30		13
8.		13	40		12
9.		13	50		14
10.		13	40		15
Sums...	242	16	0	242	14
Averages...	24°	13′	36″	24°	13′ 24″

It will at once be perceived that the average of the theodolite observations is the more accurate and reliable. *How much* more precise and reliable will be shown by a comparison of their probable errors. From the formulæ of

Art. 24, the probable error of the mean of the theodolite observations is $3''\cdot 1$, and of that of the transit $13''\cdot 8$, so that the results may be written

by the theodolite, $24° \ 13' \ 36'' \pm 3''\cdot 1$;

by the transit, $\quad 24° \ 13' \ 24'' \pm 13''\cdot 8$.

The meaning of these results is, that our confidence in the work of the theodolite is such that we could make an even wager that the *true* value of the angle is

between $24° \ 13' \ 36'' + 3''\cdot 1$ and $24° \ 13' \ 36'' - 3''\cdot 1$,

but that as far as the transit work is concerned we could only afford to bet one against one that it is

between $24° \ 13' \ 24'' + 13''\cdot 8$ and $24° \ 13' \ 24'' - 13''\cdot 8$.

The range of probable error in the first is only about one-fourth of that in the second, and hence we recognise that the theodolite average is four times as precise as that of the transit. Also, if these ten observations were to be again repeated, with equal carefulness, it will be an even wager that the new averages will differ from the true value of the angle by the respective quantities $3''\cdot 1$ and $13''\cdot 8$.

In like manner we may find that the probable error of a *single* observation is by the theodolite $9''\cdot 7$, and by the transit $43''\cdot 5$, and these observations signify that before the observations were taken, errors greater and less than those quantities were equally probable. That this is sustained in practice the reader may assure himself by assuming the averages in the above example as the true values of the angle and then computing the errors, and he will find in the one case 4 errors greater and 6 less than $9''\cdot 7$, and in the other 4 greater and 6 less than $43''\cdot 5$; which, considering that this is not an actual example but one written down at random, is a remarkably close agreement.

The probable error then furnishes us with the means of *comparing* observations. The smaller that quantity the better and more reliable are the measurements. Since the probable error r is inversely proportional to the measure of precision h (Art. 16), we see that if the probable error of one set of measurements is one-half that of a second set, its precision is twice as great; a glance at the curves of Fig. 2 shows this graphically to the eye. The probable error gives moreover an absolute measure of our confidence in the accuracy of the work.

Kinds of Observations.

18. *Direct observations* are those which are made directly upon the quantity whose magnitude is to be determined. Such are, measurements of a line by direct chaining, of an angle by direct reading with a transit, etc. They occur in the daily practice of every engineer.

19. *Indirect observations* are not made upon the quantity whose size is to be measured, but upon some other quantity or quantities related to it. Such are, measurements of a line through a triangulation by means of a base and observed angles, of an angle by regarding it as the sum or difference of other angles, the determination of the difference of level of two points by readings upon graduated rods set up at different places, the determination of latitude by observing the altitude of stars, etc. In fact the majority of observations in engineering and physical science generally belong to this class.

20. *Conditioned observations* may be either direct or indirect, but are subject to some rigorous requirement or condition. As such may be mentioned: the three measured angles in a plane triangle must be so adjusted that their sum shall be exactly $180°$, the sum of all the percentages in a chemical analysis must equal 100, the sum of the northings must equal the sum of the southings in any traverse which begins and ends at the same point, etc.

21. Measurements which are subject to no such rigorous conditions are called *independent*, meaning thereby that the observed quantities have no mutual dependence, so that all systems of values are *in thought* equally possible, and a variation of the value of one quantity need not necessarily affect the values of the others. In conditioned observations, however, all systems of values are not in thought equally possible, but only those can be admitted which exactly satisfy the rigorous conditions.

INTRODUCTION. 23

As an illustration of these classes let us consider the angles AOB and BOC having their vertices at the same point O (Fig. 3). If we set up a transit at the point O, and measure the angle AOB or BOC, each of these measurements is a *direct* observation. If however in order to deter-

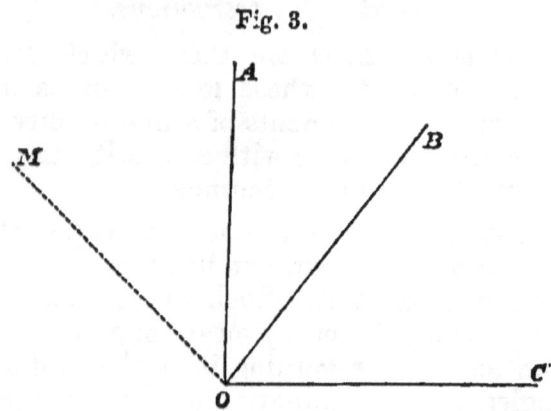

Fig. 3.

mine these angles we clamp the instrument so that its zero point has some arbitary direction OM, and then measure the angles MOA, MOB and MOC, the observations are *indirect*. Moreover, whether observed directly or indirectly, the values obtained for AOB and BOC are *independent* of each other. But if we measure the angles AOB, BOC and AOC by either of the above methods, these observations are *conditioned*, or subject to the rigorous requirement that when finally adjusted AOB plus BOC must equal AOC, and no system of values can be adopted for these three angles which does not exactly satisfy this condition.

These kinds of observations suggest the division of our subject into three parts, and we shall accordingly treat in Chapter II. of the adjustment and comparison of direct observations upon a *single* quantity; in Chapter III. of independent observations, either direct or indirect, upon *several* quantities; and in Chapter IV. of conditioned observations.

CHAPTER II.

DIRECT OBSERVATIONS UPON A SINGLE QUANTITY.

22. THE principle of the arithmetical mean has been from a very remote antiquity employed to obtain the adjusted value of an unknown quantity, which is the object of direct measurement. Its universal acceptance as an axiomatic rule for adjusting such observations, and the simplicity of the process itself, would seem to require that it should be placed at the foundation of any system for the combination of numerical measures. We must however warn the reader that the principle of the average is *not* applicable to any observations except those of equally good direct measurements upon a *single* quantity, and that it cannot be used for the combination of either direct or indirect observations upon several related quantities. A single illustration will show the danger of its indiscriminate use, and satisfy the reader that care is required to limit it to its proper sphere.

Let A and B be two points whose elevations above a given datum O are to be determined, and suppose, in order that the observations may be equally good, that these three points are situated at the vertices of an equilateral triangle. Now let a level be set up between O and A and measurements taken which show A to be 10 feet higher than O, then let it be moved to a point between O and B, and readings taken which show B to be 15 feet higher than O; and lastly let it be set between A and B, and readings taken showing B to be 4 feet above A.

These observations, it will be at once seen, are discordant, and we ask, how shall they be adjusted? Many persons would, we are afraid, proceed to apply the law of the average thus: for the elevation of A we have by one route 10 feet, and by the other 11 feet, and hence A must be taken as the average $10\frac{1}{2}$ feet; also by one route B is 15, and by the other 14 feet, and hence B should be called $14\frac{1}{2}$ feet. This however is entirely wrong, as a moment's consideration will show, for by this adjustment the measurement between O and A has been corrected from 10 to $10\frac{1}{2}$, that between O and B has been corrected from 15 to $14\frac{1}{2}$, while that between A and B remains $14\frac{1}{2} - 10\frac{1}{2} = 4$, or *has not been corrected at all*. The principle of the average does not therefore apply to such observations, or in fact to any except *equally good observations upon a single quantity*. The proper adjusted heights of A and B, we may mention, are to be determined by the methods of the next chapter and are $10\frac{1}{3}$ and $14\frac{2}{3}$ feet respectively, results which suppose the correction of each observation by an equal amount.

The Arithmetical Mean.

23. The average or arithmetical mean has always been accepted and used as the best rule for combining direct observations of equal precision upon one and the same quantity. The following reasoning may be given to justify its use.

Suppose a certain angle is to be measured, and let its true but unknown value be z. A single reading taken by a transit gives the value M_1. This being the only measurement, its value M_1 must be accepted as the most probable result, and used instead of the true value z. Now let a second reading be taken with the same instrument and under the same circumstances, which gives the value M_2. Then since there is no reason for preferring one result to the other, we consider that errors in excess and deficiency from the true value are equally probable (Art. 10), and hence combine the results so that the differences $z - M_1$ and $z - M_2$ are numerically equal, and this gives

$$z - M_1 = M_2 - z, \text{ or } z = \frac{M_1 + M_2}{2};$$

and this value of z we must accept as the most probable and must use it as representing, as far as our observations go, the true value.

In like manner if we have n observations giving the values $M_1, M_2, M_3 \ldots M_n$, errors greater and less than the true value are equally probable, and in a large number of observations will be equally numerous, and the algebraic sum of all such errors should be zero, or

$$(z - M_1) + (z - M_2) + (z - M_3) + \ldots + (z - M_n) = 0.$$

The solution of this equation will give us, as far as our n measurements are concerned, the value of z, but those measurements being limited in number and discordant, we cannot regard it as absolutely true, but only as probably true, and we call it the *most probable* value. Designating it by z_0 we have from the above equation

$$(29) \quad z_0 = \frac{M_1 + M_2 + M_3 + \ldots + M_n}{n}.$$

Hence *in any series of equally precise observations upon a single quantity, the arithmetical mean is the most probable value*, that is, the most accurate value deducible from those observations. See Art. 66.

Problem. The bearing of a line is taken five times with a solar compass giving the values,

N. 12′ E., N. 7′ E., N. 2′ W., N. 12′ W., N. 5′ E.;

what is the adjusted bearing?

$$Ans. \quad \frac{12 + 7 - 2 - 12 + 5}{5} = 2 \text{ or N. 2′ E.}$$

Probable Error of a Single Observation and of the Arithmetical Mean.

24. Let one and the same quantity, for example an angle, be measured n times with equal care; let z_0 be the average or arithmetical mean of the measurements

$$M_1, M_2, M_3, \text{ etc.}$$

Let each measurement be subtracted from the average z_0, and let the differences or *residuals* be called v_1, v_2, v_3, etc., so that

$$(31) \quad v_1 = z_0 - M_1, \quad v_2 = z_0 - M_2, \quad v_3 = z_0 - M_3, \text{ etc.}$$

DIRECT OBSERVATIONS UPON A SINGLE QUANTITY. 27

Also let Σv^2 denote the sum of the squares of these residuals, or
$$\Sigma v^2 = v_1^2 + v_2^2 + v_3^2 + \ldots + v_n^2.$$

Then, as proved in Part II., the probable error of a single observation is

$$(42) \qquad r = 0{\cdot}6745 \sqrt{\frac{\Sigma v^2}{n-1}},$$

and the probable error of the arithmetical mean z_0 is

$$(47) \qquad r_0 = \frac{r}{\sqrt{n}} = 0{\cdot}6745 \sqrt{\frac{\Sigma v^2}{n(n-1)}}.$$

To find then the probable errors r and r_0, we have to find the average, subtract from it the several measurements thus forming the residuals, then to square each of these and add the products, and then to use the above formulæ.

The reader should particularly observe the distinction between the *residual v* and the *error x* (Arts. 11, 14); the former is the difference between the *most probable* result and a single observation; the latter is the difference between the *true* value and an observation, and of course can never be exactly known since we can never be certain of having found the true value. Thus if z be the true value of an angle, and z_0 the average of the observations M_1, M_2, M_3, etc., the errors are $(z - M_1)$, $(z - M_2)$, etc., while the residuals are $(z_0 - M_1)$, $(z_0 - M_2)$, etc.

25. The probable error gives us the means of comparing the precision of our observations, and of determining the degree of confidence which we can place in them. An illustration of its use has already been given in Art. 17. We append another example worked out in full.

In the *U. S. Coast Survey Report* for 1854, the following 24 measurements are given of the angle Quaker-Pocasset-Beacon, taken at the station Pocasset in Mass., each observation being the result of an equal number of repetitions and of the same precision or weight.

We first determine the average or the most probable value of the angle by adding the readings and dividing the

28   DIRECT OBSERVATIONS UPON A SINGLE QUANTITY.

um by 24. This is 116° 43′ 49″·64. Then subtracting the
first reading from this we have the residual $v_1 = 5\cdot19$ which

No.	Reading.	v.	v^2.
	° ′ ″		
1.	116 . 43 . 44·45	5·19	26·94
2.	50·55	−0·91	·83
3.	50·95	−1·31	1·72
4.	48·90	0·74	·55
5.	49·20	0·44	·19
6.	48·85	0·79	·63
7.	47·40	2·24	5·02
8.	47·75	1·89	3·57
9.	51·05	−1·41	2·00
10.	47·85	1·79	3·20
11.	50·60	−0·96	·92
12.	48·45	1·19	1·42
13.	51·75	−2·11	4·45
14.	49·00	0·64	·41
15.	52·35	−2·71	7·34
16.	51·30	−1·66	2·75
17.	51·05	−1·41	2·00
18.	51·70	−2·06	4·24
19.	49·05	0·59	·35
20.	50·55	−0·91	·83
21.	49·25	0·39	·15
22.	46·75	2·89	8·35
23.	49·25	0·39	·15
24.	53·40	−3·76	14·14
Mean $z_0 = 116°$. 43′ . 49″·64			$\Sigma v^2 = 92\cdot15$

we place in the column headed v: and in like manner taking
the difference between the mean and each reading we fill out
that column with the values v_1, v_2, v_3, etc. Then by a table
of squares we form the numbers $v_1^2 = 26\cdot94$, $v_2^2 = 0\cdot83$, etc.,
which we place in the column v^2. Adding these we have the
sum $\Sigma v^2 = 92\cdot15$. Then as $n = 24$ we have, from the preceding formulæ, the probable error of a single observation, or

$$r = 0\cdot6745 \sqrt{\frac{92\cdot15}{23}} = 1''\cdot349,$$

and the probable error of the mean z_0 is

$$r_0 = 0.6745 \sqrt{\frac{92.15}{24 \times 23}} = 0''\cdot 275.$$

Hence the adjusted value of the angle may be written

$$116°\ 43'\ 49''\cdot 64 \pm 0''\cdot 275.$$

If then these 24 observations were to be repeated under the same circumstances, it would be an even wager that there would be 12 errors greater and 12 less that $1''\cdot 349$, and also an even wager that the mean would differ from the true value of the angle by $0''\cdot 275$. Our confidence then in the above mean $116°\ 43'\ 49''\cdot 64$ is such that we regard $0''\cdot 275$ as the error to which it is liable, that is, it is an even wager that the mean is within $0''\cdot 275$ of the true value, and of course also an even wager that it exceeds the true value by that amount.

26. From the above values given for the probable errors we observe that

$$r : r_0 :: 1 : \frac{1}{\sqrt{n}} \text{ or } r_0 = \frac{r}{\sqrt{n}}.$$

If we denote the measures of precision corresponding to r and r_0 by h and h_0, we have (Art. 16)

$$(28) \quad r = \frac{0.4769}{h} \text{ and } r_0 = \frac{0.4769}{h_0};$$

and inserting these in the expression above, we find

$$(46) \quad h_0 = h\sqrt{n},$$

that is, *the precision of the arithmetical mean increases as the square root of the number of observations.*

In order then to make the value of the angle in the above examples twice as precise, that is, make the probable error of the mean one-half as large, we must have four times as many observations, or 96. Let the reader test this rule by taking at random any six of those observations, and find the probable errors. The probable error of the mean will be approximately twice that given above, or

$$2 \times 0''\cdot 275 = 0''\cdot 55,$$

while the probable error of a single observation will remain nearly the same as before, or $1''\!\cdot\!35$.

Problems. 1. A base line is measured five times with a steel tape reading to hundredths of a foot, and also five times with a chain reading to tenths of a foot, with the following results:

By the tape.	By the chain.
741·17 feet	741·2 feet
741·09 ...	741·4 ...
741·22 ...	741·0 ...
741·12 ...	741·3 ...
741·10 ...	741·1 ...

What are the averages and their probable errors?

Ans. By tape, $741\!\cdot\!14 \pm 0\!\cdot\!016$.

2. A line is measured five times, and the probable error of the mean is 0·016 feet. How many additional measurements of the same precision are necessary in order that the probable error of the mean shall be only 0·004 feet?

Weights of Observations. The General Mean.

27. We have thus far considered our observations as equally precise, that is, as made with the same instrument, and under exactly the same circumstances. We now come to the case of observations of unequal precision, or as we usually say of unequal *weight*. The sense in which we use this word we will try to make clear by a practical illustration. Suppose a line to be measured 20 times with the *same* chain, 10 measurements giving the value 934·2 feet, 7 giving the value 934·0 feet, and 3 giving 934·4. The adjustment of these results is effected by the above method of the average, by writing the first value 10 times, the second 7 and the third 3, adding the numbers and dividing their sum by 20; this process is evidently the same as multiplying each value by the number of times it occurs, and dividing the sum of the three products by 20, or

$$\frac{10 \times 934\!\cdot\!2 + 7 \times 934\!\cdot\!0 + 3 \times 934\!\cdot\!4}{20}$$

$$= 934\!\cdot\!0 + \frac{10 \times 0\!\cdot\!2 + 3 \times 0\!\cdot\!4}{20} = 934\!\cdot\!16.$$

DIRECT OBSERVATIONS UPON A SINGLE QUANTITY. 31

Now if the 10 measurements, instead of giving each time 934·2 feet, had given 10 results whose mean was that number, 10 would be the *weight* of 934·2, or considering 934·2 as a single observation, 10 is said to be its weight. Also 7, 3, and 20 are the *weights* of 934·0, 934·4, and 934·16: the first being the equivalent of 7, the second of 3, and the third of 20 observations, each equally good.

By the term weights, then, we mean numbers related to the accuracy of the measurements, so that an observation of the weight 8 is to be regarded as equal to 8 observations of the weight 1; and observations having the same weight are to be considered as equally good. The average of n equally good observations, being the equivalent of those observations, has thus a weight of n.

The combination of weighted observations is to be made by the process illustrated by the above numerical example; thus if g_1, g_2, g_3, etc. be the weights of the measurements M_1, M_2, M_3, etc., the adjusted value is

$$(50) \quad Z = \frac{g_1 M_1 + g_2 M_2 + g_3 M_3 + \text{etc.}}{g_1 + g_2 + g_3 + \text{etc.}},$$

or, *the most probable value of the measured quantity is found by multiplying each observation by its weight, and dividing the sum of the products by the sum of the weights.* This value Z is called the *General Mean* to distinguish it from the arithmetical mean z_0, which is only to be used when all the weights are equal.

Weights should be carefully distinguished from measures of precision (Art. 16): the former are *relative numbers*, which are usually so taken as to be free from fractions; the latter are *absolute quantities*. The relation between them will be shown in the next article.

Problems. 1. An angle is measured 20 times with the same theodolite. The mean of 6 readings is 27° 34′ 32″, the mean of 10 is 27° 34′ 40″, and of the other 4 is 27° 34′ 48″. What is the adjusted value? *Ans.* 27° 34′ 39″·2.

2. The bearing of a certain line is taken as N. 89° 45′ W., and as S. 89° 45′ W. If the weight of the first observation

is 13, and that of the second 2, what is the most probable value of the bearing? *Ans.* N. 89° 49′ W. whose weight is 15.

28. To establish the relation between weights and probable errors, let us consider n observations upon the same quantity, giving the results M_1, M_2, M_3, etc. whose measures of precision are h_1, h_2, h_3, etc. If z is the *true* value of the quantity, the errors are $(z - M_1) = x_1$, $(z - M_2) = x_2$, etc. Then from Art. 11,

$$\text{Prob. of the error } x_1 = y_1 = c_1 e^{-h_1^2 x_1^2},$$
$$\ldots\ldots\ldots\ldots\ldots\ldots\; x_2 = y_2 = c_2 e^{-h_2^2 x_2^2},$$
$$\ldots\ldots\ldots\ldots\ldots\ldots\; x_3 = y_3 = c_3 e^{-h_3^2 x_3^2},$$
$$\text{etc.,} \qquad \text{etc.}$$

From Art. 8 the probability P of committing all these errors is the product of the probabilities y_1, y_2, y_3, etc., or

$$(25) \qquad P = c_1 c_2 c_3 \ldots e^{-(h_1^2 x_1^2 + h_2^2 x_2^2 + h_3^2 x_3^2 + \text{etc.})};$$

and the most probable value of z is that for which P is a maximum (Art. 9), and in order that P should be a maximum, the quantity,

$$(26) \qquad h_1^2 x_1^2 + h_2^2 x_2^2 + h_3^2 x_3^2 + \text{etc.,}$$

must be a minimum. Inserting in this the values of x_1, x_2, x_3, etc. in terms of z, we have

$$h_1^2 (z - M_1)^2 + h_2^2 (z - M_2)^2 + h_3^2 (z - M_3)^2 + \text{etc.} = \text{a minimum.}$$

Differentiating it with reference to z, dividing by $2dz$ and placing the derivative equal to zero, we have

$$h_1^2 (z - M_1) + h_2^2 (z - M_2) + h_3^2 (z - M_3) + \text{etc.} = 0.$$

Solving this equation, and denoting the resulting value of z by Z, we find the most probable result to be

$$(48) \qquad Z = \frac{h_1^2 M_1 + h_2^2 M_2 + h_3^2 M_3 + \text{etc.}}{h_1^2 + h_2^2 + h_3^2 + \text{etc.}}.$$

The value of Z given by this expression must be the same as that given by the general mean in Art. 27. Comparing the two expressions, we see that

$$g_1 : g_2 : g_3 :: h_1^2 : h_2^2 : h_3^2,$$

or, *the weights of observations are proportional to the squares of their measures of precision.*

A direct application of these principles can be made to the very common case of different sets of observations of unequal precision, arising from measurements by different instruments of one and the same quantity. Let the number of observations in the first series be n_1, in the second n_2, in the third n_3, etc.; let the mean as given by the first series be z_1, by the second z_2, by the third z_3, etc.; and let the weights of these be g_1, g_2, g_3, etc.; and the corresponding measures of precision be h_1, h_2, h_3, etc. Then from the above principle

$$(49) \quad g_1 : g_2 : g_3 :: h_1^2 : h_2^2 : h_3^2.$$

Inserting in these the values of h_1, h_2, etc., which are determined in Part II., we have (after striking out the common factor 2)

$$(52) \quad g_1 : g_2 : g_3 :: \frac{n_1(n_1-1)}{\Sigma v'^2} : \frac{n_2(n_2-1)}{\Sigma v''^2} : \frac{n_3(n_3-1)}{\Sigma v'''^2},$$

in which $\Sigma v'^2$ denotes the sum of the squares of the residuals in the first set, $\Sigma v''^2$ in the second, etc. (Art. 24). We have then in such a case to find the average of each set, next the weights of these averages by the proportion just given, and then the most probable value of the measured quantity by the general mean (Art. 27) of the several arithmetical means, that is, by the formula

$$(50) \quad Z = \frac{g_1 z_1 + g_2 z_2 + g_3 z_3 + \text{etc.}}{g_1 + g_2 + g_3 + \text{etc.}}.$$

To illustrate this very common case we give the following example and problems.

A certain line was measured by three different surveying parties, using three different chains; the first party measured it 5 times with the results given in column I. below, the second 6 times with results as in column II., and the third 4 times as shown in column III. What are the relative weights of the three means and the most probable length of the line?

We first determine the three means z_1, z_2, and z_3: then subtracting each observation from its mean, find the residuals

I.	v'.	v'^2.	II.	v''.	v''^2.	III.	v'''.	v'''^2.
5110	2	4	4980	120	14400	5105	0	0
5090	22	484	5100	0	0	5100	5	25
5140	−28	784	5220	−120	14400	5110	−5	25
5100	12	144	5160	−60	3600	5105	0	0
5120	−8	64	5040	60	3600			
			5100	0	0			
$z_1 = 5112$	$\Sigma v'^2 = 1480$		$z_2 = 5100$	$\Sigma v''^2 = 36000$		$z_3 = 5105$	$\Sigma v'''^2 = 50$	

which are placed in the columns v', v'', and v'''. From a table of squares we take the squares of these numbers and place them in the columns headed v'^2, v''^2, and v'''^2: and by their addition find the sums of those squares. For the first series of measurements $n_1 = 5$, for the second $n_2 = 6$, and for the third $n_3 = 4$. Designating the weights of the means by g_1, g_2, and g_3, we have from the above proportion

$$g_1 : g_2 : g_3 :: \frac{5 \times 4}{1480} : \frac{6 \times 5}{36000} : \frac{4 \times 3}{50},$$

or by reduction

$$g_1 : g_2 : g_3 :: 600 : 37 : 10656;$$

hence the mean z_1 is equivalent to 600 observations of the weight 1, z_2 to 37, and z_3 to 10656; the second is hence the least, and the third the most reliable. The most probable value of the line then is

$$Z = \frac{600 \times 5112 + 37 \times 5100 + 10656 \times 5105}{600 + 37 + 10656} = 5105 \cdot 36.$$

Problems. 1. An angle is measured four times with a theodolite, six times with a transit, and five times with a sextant, giving the observations:

DIRECT OBSERVATIONS UPON A SINGLE QUANTITY.

By the theodolite.	By the transit.	By the sextant.
6° 17′ 5″	6° 17′	6° 17′ 20″
6 17 10	6 16	6 17 0
6 17 0	6 15	6 17 40
6 17 5	6 19	6 16 50
	6 17	6 17 10
	6 18	

What are the relative weights of the means, and the most probable value of the angle?

Ans. $Z = 6°\ 17'\ 5''\cdot 36$.

2. Two sets of measurements are made upon the same angle by different observers with the same instrument. Each takes 5 observations, as follows:

First Observer.	Second Observer.
47° 23′ 40″	47° 23′ 30″
23 45	23 40
23 30	23 50
23 35	24 0
23 40	23 20

What are the relative weights of the observers, and the adjusted value of the angle?

Ans. $g_1 : g_2 :: 100 : 13$.

29. Observations may also be combined when their probable errors are known. From Art. 28 we have the proportion

$$(49) \quad g_1 : g_2 : g_3 :: h_1^2 : h_2^2 : h_3^2,$$

and also from Art. 16 the relation

$$(28) \quad h_1^2 : h_2^2 : h_3^2 :: \frac{1}{r_1^2} : \frac{1}{r_2^2} : \frac{1}{r_3^2};$$

hence by comparison of the two proportions

$$(53) \quad g_1 : g_2 : g_3 :: \frac{1}{r_1^2} : \frac{1}{r_2^2} : \frac{1}{r_3^2};$$

that is, *the weights of observations are inversely proportional to the squares of their probable errors.* Hence having computed the probable errors of two arithmetical means, the relative weights can be immediately found, and the two means then combined by the general mean.

Thus in Art. 17 we have a case where the same angle is measured by a theodolite and by a transit, giving the means and probable errors,

by the theodolite, $24° \ 13' \ 36'' \pm 3''\cdot 1$;

by the transit, $\quad 24° \ 13' \ 24'' \pm 13''\cdot 8$.

Designating the weights of these by g_1 and g_2, we have from the above principle,

$$g_1 : g_2 :: \frac{1}{3\cdot 1^2} : \frac{1}{13\cdot 8^2} :: 119 : 6 \text{ nearly};$$

and the most probable value of the angle then is

$$Z = 24° \ 13' + \frac{119 \times 36'' + 6 \times 24''}{125} = 24° \ 13' \ 35''\cdot 4.$$

Problems. 1. What is the most probable value of the base line in Prob. 1 of Art. 26, as given by the two sets of observations? *Ans.* 741·146.

2. Three sets of observations upon the same quantity give the following averages and probable errors,

$$803\cdot 4 \pm 0\cdot 4 ; \quad 803\cdot 2 \pm 0\cdot 3 ; \quad 803\cdot 1 \pm 0\cdot 08.$$

What is the adjusted value of the quantity?

Probable Error of the General Mean.

30. Having determined the adjusted value of observations of unequal weights by the preceding methods, we next inquire what is our degree of confidence in that result, or what is its probable error (Art. 16).

Let n be the number of observations, or sets of measurements, g_1, g_2, g_3, etc. their relative weights; let Z be the general mean as found by Art. 27, and G its weight. Also let v_1, v_2, v_3, etc. be the residuals or differences between the general mean and each observation, and Σgv^2 the sum

$g_1v_1^2 + g_2v_2^2 + g_3v_3^2 +$ etc., that is, the sum of the quantities formed by multiplying the square of each residual by its corresponding weight. Then, as proved in Part II., the probable error of an observation *whose weight is unity* is

$$(65) \qquad r = 0{\cdot}6745 \sqrt{\frac{\Sigma g v^2}{n-1}}.$$

Having found r, the probable error of the general mean, or of any observation, may be found by the principle of Art. 29. If G be the weight of the general mean and R its probable error, g_1 the weight of a given observation and r_1 its probable error, we have

$$(53) \qquad G : g_1 : 1 :: \frac{1}{R^2} : \frac{1}{r_1^2} : \frac{1}{r^2},$$

from which we find

$$(56) \qquad R = \frac{r}{\sqrt{G}} \text{ and } r_1 = \frac{r}{\sqrt{g_1}};$$

that is, *the probable error of an observation whose weight is known, is equal to the probable error of an observation of the weight 1, divided by the square root of the given weight.* The weight G of the general mean is always equal to

$$g_1 + g_2 + g_3 + \text{ etc. (Art. 27).}$$

If we have then several sets of measurements as in Art. 28, whose averages are z_1, z_2, z_3, etc., we have to find their relative weights g_1, g_2, g_3, etc., and their general mean Z. Subtracting each average from the general mean gives us the residuals v_1, v_2, etc.; multiplying the square of each by its weight and adding the products gives us the sum $\Sigma g v^2$. Then n being the number of sets, the above formulæ furnish us with the probable errors. Further, if we have found the probable errors r_1, r_2, etc. of the averages z_1, z_2, etc., the principle of Art. 29 gives us for the probable error of the general mean

$$(54) \qquad R = r_1 \sqrt{\frac{g_1}{G}} = r_2 \sqrt{\frac{g_2}{G}} = \text{etc.,}$$

which may in some cases be more convenient to use than the formulæ above. The following examples and problems will illustrate the application of the formulæ.

38 DIRECT OBSERVATIONS UPON A SINGLE QUANTITY.

1. Suppose that the observations in the example of Art. 25 are given as in the following column z, the mean of the first five being 48″·81 with the weight 5, the mean of the next four 48″·76 with the weight 4, and so on. Then the operation for finding the probable error of the general mean is thus exhibited:

g.	z.	v.	v^2.	gv^2.
	° ′ ″			
5	116 . 43 . 48·81	0·83	0·69	3·45
4	48·76	0·88	0·77	3·08
5	49·53	0·11	0·01	0·05
3	51·56	−1·92	3·69	11·07
2	50·38	−0·74	0·55	1·10
5	49·84	−0·20	0·04	0·20
$G = 24$	$Z = 116° \ 43' \ 49·64$			$\Sigma gv^2 = 18·95$

The general mean Z of course here agrees with the average found in Art. 24, and its weight G is the sum of the several weights or 24 the number of single observations. Subtracting each average from the general mean we have the residuals in the column v, and from a table of squares we place their squares in the column v^2; multiplying each of these by its corresponding weight we have the quantities in the column gv^2 whose sum is 18·95. Then, n being 6, we have from the formula

$$r = 0·6745 \sqrt{\frac{18·95}{5}} = 1''·32.$$

This is the probable error of an observation of the weight unity, and should hence agree with that of a single observation 1″·35, as found in Art. 24. The discrepancy is due to the small value of n (see Prob. 2). The probable error of the general mean is then

$$R = \frac{r}{\sqrt{G}} = \frac{1''·32}{\sqrt{24}} = 0·269,$$

which agrees sufficiently well for most practical comparisons with that found before.

2. In Prob. 1 of Art. 28 the probable error of the theodolite mean is $1'''\cdot 4$, of the transit $23'''\cdot 3$, and of the sextant $5''\cdot 8$, and their relative weights are nearly 288, 1 and 16. Hence the probable error of the general mean is by our second formula

$$R = 1\cdot 4 \sqrt{\frac{288}{305}} = 5\cdot 8 \sqrt{\frac{16}{305}} = 1''\cdot 33.$$

The first formula gives a less result, as n is only 3. Strict agreement in such results cannot be expected, since the theory upon which the formulæ are deduced supposes n to be a large number, and in our examples illustrating their application we are obliged to choose cases involving but few observations.

Problems. 3. What is the general mean of the observations on the base line in the question of Art. 26, and its probable error? *Ans.* $741\cdot 146 \pm 0\cdot 012$.

4. Eight observations of a quantity give the results 769, 768, 767, 766, 765, 764, 763 and 762, whose relative weights are 1, 2, 3, 4, 5, 6, 7 and 8. What is the probable error of the general mean, and the probable error of each observation?

Recapitulation.

31. We have now given and illustrated the methods for adjusting and comparing direct observations upon a single quantity. They fall under three heads:

1st. If all the observations are equally good or of equal weight, the average is the most probable result (Art. 23). The degree of confidence which we can place in one of those observations, or in the average, is shown by their probable errors (Art. 24).

2nd. If there are several sets of observations made under different circumstances or by different instruments, the most probable result is given by the general mean (Art. 27), and in order to obtain that result the relative weights of the several averages must be deduced by the proportion of Art. 28. The precision of the several means and of the final general mean is also shown by their probable errors (Art. 30).

3rd. If we have single observations which are known to be of unequal precision, we have no means of finding their weights as in the preceding case, but must assign to them such weights as they seem to deserve in our judgment. Thus if an angle be measured once by a theodolite reading to 20", and once by a transit reading to 1', we recognise that the first is the more reliable, and should, in finding the mean, give it a weight of about 9 times as much as the second. By taking series of observations with two such instruments, their relative weights may be found and recorded for use in cases where single measurements arise. When such knowledge does not exist, weights may be assigned, and then the general mean found as in Art. 27. The assigning of weights in such case is of course a matter requiring experience and judgment.

The combination of direct observations upon *different* but related quantities is considered in the two following chapters. The most simple cases of such adjustments, such as finding the length of a base line, which has been measured in several portions, are referred to in Arts. 41 and 42, in connection with the discussion of their probable errors.

CHAPTER III.

INDEPENDENT OBSERVATIONS UPON SEVERAL QUANTITIES.

32. IN Arts. 18 and 19 we divided observations into direct and indirect, the former being made upon the quantities to be determined, and the latter upon other quantities related to them. The line of demarcation between the two is however not very distinct, nor is it necessary that we should be able to point it out clearly. For practical purposes independent observations need only to be distinguished as those upon *one* quantity, which have been already considered, and as those involving *more than one* quantity, which we are now to take up. The methods and formulæ of the preceding chapter which treat of one quantity, although for convenience given separately, are indeed but particular cases of those now to be developed.

Independent observations, whether direct or indirect, which are made to determine the magnitudes of quantities, are generally represented by equations which we shall call *observation equations*. To illustrate how they arise let us consider the following practical case. Let O (Fig. 4) represent a given bench-mark, and S, T, U three points whose elevations above O are to be determined. Let five lines of levels be run between these points as indicated by the dotted lines of the figure, giving the following results:

Fig. 4.

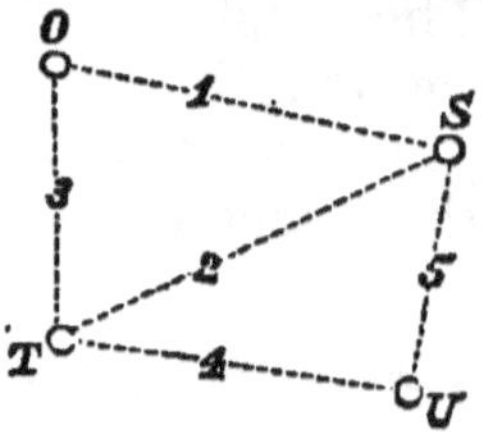

Observation 1. S above $O = 10$ feet,
.... 2. T ...... $S = 7$ ...
.... 3. T ...... $O = 18$ ...
.... 4. T ...... $U = 9$ ...
.... 5. U below $S = 2$ ...

It will be at once perceived that the measurements are discordant; if we take observations 1 and 2 as correct, the height of S is 10 feet and T is 17; if 2 and 3 are correct, S is 11 and T is 18 feet, etc.: and in general it will be found impossible to find a system of values which will *exactly* satisfy all the observations. If we designate the elevations of the points S, T and U by the letters s, t and u, the observations furnish the following equations:

$$s = 10,$$
$$t - s = 7,$$
$$t = 18,$$
$$t - u = 9,$$
$$s - u = 2,$$

each one of which is an approximation to the truth, but all of which cannot be correct. The number of these equations is 5, the number of the unknown quantities is 3, and hence an exact solution cannot be made by algebraic processes. It being impossible then to find values of s, t, and u which will satisfy all the equations, we must be content with determining their *most probable* values (Art. 9).

So, in general, independent observations upon several quantities give rise to independent *observation equations* greater in number than the unknown quantities to be determined; and our problem is, to find, out of the many systems of values, all equally *possible*, which may be assigned to the unknown quantities, a system which is the *most probable*, and hence the best.

Solution of Observation Equations.

33. We take up first the case of measurements of equal precision or of equal weight. Let M_1, M_2, M_3, etc. be the numerical results of the observations which are n in number, and which are made upon quantities related to the quantities s, t, u, etc., whose values are to be found. Let the observations give rise to the following observation equations:

$$\begin{aligned} a_1 s + b_1 t + c_1 u + \text{etc.} &= M_1, \\ a_2 s + b_2 t + c_2 u + \text{etc.} &= M_2, \\ a_3 s + b_3 t + c_3 u + \text{etc.} &= M_3, \\ \text{etc.,} \quad \text{etc.,} & \end{aligned} \quad (66)$$

UPON SEVERAL QUANTITIES. 43

in which a_1, a_2, b_1, b_2, etc. are the known coefficients of the unknown quantities. The number of the measurements is n, and if the number of unknown quantities were also n the solution of the equations could at once be made. But as we have seen the number of the latter is usually less than n, and an exact solution is impossible. We must therefore be content with finding the most probable values of s, t, u, etc.

Whatever system of values is chosen for s, t, u, etc. it will not exactly satisfy each of the above equations, since the measurements M_1, M_2, M_3, etc. are imperfect. If then we consider s, t, u, etc. as representing the *true* values of the quantities, the above equations may be written

$$(67) \quad \begin{aligned} a_1 s + b_1 t + c_1 u + \text{etc.} - M_1 &= x_1, \\ a_2 s + b_2 t + c_2 u + \text{etc.} - M_2 &= x_2, \\ a_3 s + b_3 t + c_3 u + \text{etc.} - M_3 &= x_3, \\ \text{etc.,} \quad \text{etc.,} \end{aligned}$$

that is to say, they do not reduce exactly to zero for any values of the unknown quantities, but leave small differences or errors x_1, x_2, x_3, etc. Now by our general principle of Art. 14 the most probable values of the unknown quantities are those which make the sum of the squares of the errors a minimum, and hence the most probable values of s, t, u, etc. are those which make $x_1^2 + x_2^2 + x_3^2 + $ etc. the least possible. We procede to develope a method for finding those quantities.

Let us first consider what is the most probable value of the unknown quantity s. As we need only to regard s, let us denote the terms in the above equations, independent of s, by the letters N_1, N_2, N_3, etc. Then they become

$$\begin{aligned} a_1 s + N_1 &= x_1, \\ a_2 s + N_2 &= x_2, \\ a_3 s + N_3 &= x_3, \\ \text{etc.} \end{aligned}$$

Squaring both terms of each of these equations and adding the results, we have

$$(a_1 s + N_1)^2 + (a_2 s + N_2)^2 + (a_3 s + N_3)^2 + \text{etc.} = x_1^2 + x_2^2 + x_3^2 + \text{etc.}$$

According to the principle above stated this quantity is to be made a minimum to give the most probable value of s.

Differentiating it with respect to s, placing the first differential coefficient equal to zero, and dividing by 2, we have

$$a_1(a_1 s + N_1) + a_2(a_2 s + N_2) + a_3(a_3 s + N_3) + \text{etc.} = 0,$$

and this is the equation which furnishes us with the most probable value of s. Hence we have the principle: *To form the equation which gives the most probable value for one of the unknown quantities as s, we multiply each of the observation equations by the coefficient of s in that equation and add the results.* In the same way, to find the equation for t we multiply each observation equation by the coefficient of t in that equation and add the results. The equations thus formed are called *normal equations;* they will be the same in number as the number of the unknown quantities, and will be satisfied by only one system of values of the unknown quantities, which will therefore be the most probable system.

34. For illustration, let us suppose that four measurements upon three unknown quantities have given the observation equations

$$s - t + 2u = 3 \quad \ldots\ldots\ldots\ldots (1),$$
$$3s + 2t - 5u = 5 \quad \ldots\ldots\ldots\ldots (2),$$
$$4s + t + 4u = 21 \quad \ldots\ldots\ldots\ldots (3),$$
$$-s + 3t + 3u = 14 \quad \ldots\ldots\ldots\ldots (4),$$

from which it is required to find the best system of values of s, t, and u. To form the normal equation for s, we must multiply each equation by the coefficient of s in that equation and add the results; hence multiplying equation (1) by 1, equation (2) by 3, equation (3) by 4, and equation (4) by -1, we have

$$s - t + 2u = 3,$$
$$9s + 6t - 15u = 15,$$
$$16s + 4t + 16u = 84,$$
$$s - 3t - 3u = -14;$$

and adding these, we have the first normal equation

$$27s + 6t = 88 \quad \ldots\ldots\ldots\ldots (5).$$

We must now perform the same operation for t: multiplying equation (1) by -1, equation (2) by 2, equation (3) by 1, and equation (4) by 3, and adding the results, we have the normal equation for t, viz.

$$6s + 15t + u = 70 \quad \ldots\ldots\ldots\ldots(6).$$

In like manner, we multiply equation (1) by 2, equation (2) by -5, equation (3) by 4, equation (4) by 3, and add the results to form the normal equation for u, viz.

$$t + 54u = 107 \quad \ldots\ldots\ldots\ldots(7).$$

By this process we have three normal equations (5), (6) and (7), containing only three unknown quantities, and solving these by any of the algebraic methods, we find

$$s = 2{\cdot}4702, \quad t = 3{\cdot}5509, \quad u = 1{\cdot}9157.$$

If we substitute these values in equations (1), (2), (3) and (4) we shall find that they will not reduce to zero, but give the residuals

$$v_1 = -0{\cdot}2493, \quad v_2 = -0{\cdot}0661, \quad v_3 = 0{\cdot}0945, \quad v_4 = -0{\cdot}0704,$$

the sum of whose squares is $0{\cdot}0804$. (Note,—the student should not forget the distinction between the *errors* x_1, x_2, etc. and the *residuals* v_1, v_2, etc. See Art. 24.) This quantity $0{\cdot}0804$ is less than the sum of the squares of the residuals resulting from any other values of s, t, and u. Let the reader test this by trying other values, for instance

$$s = 2\tfrac{1}{4}, \quad t = 3\tfrac{2}{3}, \quad \text{and } u = 1\tfrac{9}{10}.$$

To solve observation equations of equal weight, we have then the following: *For each of the unknown quantities form a normal equation by multiplying each observation equation by the coefficient of that unknown quantity in that equation (taken with its proper sign), and adding the results. Then there will be as many normal equations as unknown quantities, and their solution will give the most probable values of those unknown quantities.* In forming the normal equations it should be particularly noticed that the signs of the coefficients (+ or −) are to be observed in performing the multiplications; and also that when the unknown quantity under consideration does not occur in an observation equation its coefficient is 0. For further illustration we give an additional example, and a couple of problems as exercises for the student.

If we have given the observation equations
$$s = 14,$$
$$t - s = 7,$$
$$t = 20,$$
the coefficient of s in the first is 1, in the second -1, and in the third 0. Hence multiplying each equation by these numbers and adding the results, we have
$$2s - t = 7$$
as the normal equation for s. Also the coefficient of t in the first equation is 0, in the second 1, and in the third 1, and in like manner the normal equation for t is
$$2t - s = 27.$$
Solving the two normal equations, we find as the best values
$$s = 13\tfrac{2}{3}, \quad \text{and} \quad t = 20\tfrac{1}{3}.$$

Problems. 1. Form the normal equations, and find the most probable values of s, t and u from the observation equations stated in Art. 32.

Ans. The normal equations are
$$3s - t - u = 5,$$
$$-s + 3t - u = 34,$$
$$-s - t + 2u = -11,$$
from which $s = 10\tfrac{3}{8}$, $t = 17\tfrac{5}{8}$, and $u = 8\tfrac{1}{2}$ feet.

2. Find the most probable values of x, y, and z from the observation equations
$$x = 5,$$
$$y + x = 11,$$
$$z + y + x = 13,$$
$$z + y - x = 4.$$

Ans. $x = 4\tfrac{2}{3}$, etc.

Adjustment of Independent Observations of equal weight.

35. The preceding principles furnish us with the following method of adjusting independent observations, either direct or indirect, upon several quantities.

1st. Represent each quantity to be determined by a symbol s, t, u, etc., and for each observation write an *observation equation* (Art. 32).

2nd. From the observation equations form the *normal equations* (Arts. 33, 34), which will be as many as there are unknown quantities.

3rd. Solve the normal equations by any convenient algebraic method; the resulting values of the unknown quantities will be their most probable values (Art. 33), that is, the best values which can be deduced from the given observations.

The following examples will illustrate the application of the method to cases arising in ordinary engineering practice.

I. *Adjustment of level lines.* In the *Report of the U. S. Geological and Geographical Survey of the Territories* for 1873, Mr J. T. GARDNER gives the following measurements as deduced from an examination of railroad profiles and coast survey levels.

1. S above O, 573·08 feet, by Coast Survey, and Canal levels, via Albany.
2. T „ S, 2·60 „ „ Observations on surface of Lake Erie.
3. T „ O, 575·27 „ „ Coast Survey and R. R. levels, via Albany.
4. U „ T, 167·33 „ „ R. R. levels.
5. X „ U, 3·80 „ „ „ „
6. X „ T, 170·28 „ „ „ via Alliance and Crestline.
7. X „ Y, 425·00 „ „ „ „
8. Y „ O, 319·91 „ „ R. R. and Coast Survey levels, via Phil.
9. Y „ O, 319·75 „ „ R.R. levels, via Baltimore.

In which

O is the mean surface of the Atlantic Ocean,

S is the mean surface of Lake Erie at Buffalo,
T is Cleveland city datum plane,
U is Depot track at Columbus, Ohio,
X is Union Depot track at Pittsburg,
Y is Depot track at Harrisburg.

It is required to find the most probable elevations of each of these points above the datum O, as given by these nine observations, supposed here to be of equal reliability or weight.

We represent the unknown heights of S, T, U, X and Y by the letters s, t, u, x and y. Then the observations give us the equations

$$\begin{aligned} s &= 573{\cdot}08, \\ t - s &= 2{\cdot}60, \\ t &= 575{\cdot}27, \\ u - t &= 167{\cdot}33, \\ x - u &= 3{\cdot}80, \\ x - t &= 170{\cdot}28, \\ x - y &= 425{\cdot}00, \\ y &= 319{\cdot}91, \\ y &= 319{\cdot}75. \end{aligned}$$

All the coefficients of the unknown quantities are either $+1$ or -1. Multiplying each equation in which s occurs by its coefficient in that equation, and adding the products, we form the normal equation for s; then multiplying each equation in which t occurs by its coefficient, etc., gives us the second normal equation. Thus we have the five normal equations

$$\begin{aligned} 2s - t &= 570{\cdot}48 = A, \\ -s + 4t - u - x &= 240{\cdot}26 = B, \\ -t + 2u - x &= 163{\cdot}53 = C, \\ -t - u + 3x - y &= 599{\cdot}08 = D, \\ -x + 3y &= 214{\cdot}66 = E, \end{aligned}$$

containing only five unknown quantities. By whatever process these equations be solved, the values found for s, t, u, etc.

will be the same. We leave the reader then to choose his own process, simply remarking that the method of indeterminate multipliers will prove the shortest. Representing the numerical terms by the letters A, B, C, D and E, we find

$$51s = 32A + 13B + 11C + 9D + 3E = 29213{\cdot}27,$$
$$51t = 13A + 26B + 22C + 18D + 6E = 29332{\cdot}14,$$
$$51u = 11A + 22B + 50C + 27D + 9E = 37844{\cdot}70,$$
$$17x = 3A + 6B + 9C + 12D + 4E = 12672{\cdot}37,$$
$$17y = A + 2B + 3C + 4D + 7E = 5440{\cdot}53;$$

hence we have

$s = 572{\cdot}81$, while Mr GARDNER gives $573{\cdot}08$,
$t = 575{\cdot}14$, ... $575{\cdot}68$,
$u = 742{\cdot}05$, ... $742{\cdot}60$,
$x = 745{\cdot}43$, ... $746{\cdot}00$,
$y = 320{\cdot}05$, ... $319{\cdot}91$.

The discrepancy in the results is mainly due to the fact that Mr GARDNER has considered the observations as of very unequal weight, taking s and t for instance, as given by 1 and 2 alone without reference to the other measurements (see Arts. 37 and 40).

If these values be substituted in the observation equations, the residuals and their squares will be

From our values.

No.	v.	v^2.
1.	0·27	0·073
2.	·27	·073
3.	·13	·017
4.	·42	·176
5.	·42	·176
6.	·01	·000
7.	·38	·144
8.	·14	·020
9.	·30	·090
		$\Sigma v^2 = 0{\cdot}769$

From GARDNER's values.

No.	v.	v^2.
1.	0·00	0·000
2.	·00	·000
3.	·41	·168
4.	·40	·160
5.	·40	·160
6.	·04	·002
7.	1·09	1·188
8.	·00	·000
9.	·16	·026
		$\Sigma v^2 = 1{\cdot}704$

The sum 0·769 of the squares of the residuals is the least arising from all the systems of values which can be attributed to s, t, u, etc. Mr GARDNER'S values, although exactly satisfying three of the equations, give the corresponding sum 1·704.

The above process, though simple, can be considerably abridged in the numerical operations by assuming approximate values for the heights of S, T, etc., and regarding the unknown quantities as corrections to be applied to those assumed values. We see at once from the observations that 573 and 575 feet would be approximate heights for S and T. If then we place

$$s = 573 + s',$$
$$t = 575 + t',$$
$$u = 742 + u',$$
$$x = 745 + x',$$
$$y = 320 + y',$$

in which s', t', u', etc. are corrections to be applied to the approximate elevations 573, 575, 742, etc., we may insert these expressions for s, t, u, etc. in the observation equations above and obtain

$$s' = 0·08,$$
$$t' - s' = 0·60,$$
$$t' = 0·27,$$
$$u' - t' = 0·33,$$
$$x' - u' = 0·80,$$
$$x' - t' = 0·28,$$
$$x' - y' = 0·00,$$
$$y' = -0·09,$$
$$y' = -0·25.$$

From these we form the normal equations by the same process as before, and have

$$2s' - t' = -0·52 = A',$$
$$-s' + 4t' - u' - x' = 0·26 = B',$$
$$-t' + 2u' - x' = -0·47 = C',$$
$$-t' - u' + 3x' - y' = 1·08 = D',$$
$$-x' + 3y' = -0·34 = E',$$

and their solution gives
$$51s' = 32A' + 13B' + 11C' + 9D' + 3E' = -9{\cdot}73,$$
$$51t' = 13A' + 26B' + 22C' + 18D' + 6E' = 7{\cdot}06,$$
etc., etc.,

from which
$$s' = -0{\cdot}19,\ t' = 0{\cdot}14,\ u' = 0{\cdot}05,\ x' = 0{\cdot}43,\ y' = 0{\cdot}05;$$
and adding these to the assumed approximate values, we have
$$s = 573 - 0{\cdot}19 = 572{\cdot}81,$$
$$t = 574 + 0{\cdot}14 = 575{\cdot}14,$$
$$u = 742 + 0{\cdot}05 = 742{\cdot}05,$$
$$x = 745 + 0{\cdot}43 = 745{\cdot}43,$$
$$y = 320 + 0{\cdot}05 = 320{\cdot}05,$$
which are the same as obtained by the longer method.

II. *Adjustment of Angles taken at a point.* At a station O there were measured the following horizontal angles, each being the average of an equal number of readings.

Fig. 5.

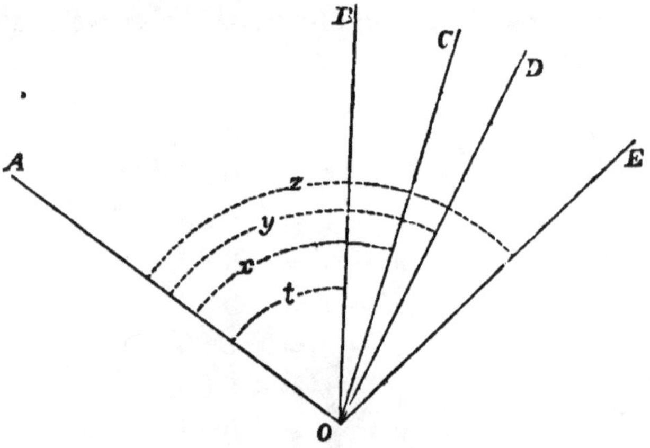

$AOB = 58° 56' 42''$,
$AOD = 76\ \ 43\ \ \ 6$,
$BOC = 13\ \ 14\ \ 15$,
$BOD = 17\ \ 46\ \ 26$,
$BOE = 34\ \ 14\ \ 17$,
$COD = \ \ 4\ \ 32\ \ \ 7$,
$DOE = 16\ \ 27\ \ 54$,

and it is required to find the most probable values of all the angles AOB, BOC, etc. In the solution of such a problem it will be found most convenient to take as the unknown quantities the direction angles $AOB = t$, $AOC = x$, $AOD = y$, and $AOE = z$. Then the observation equations will be

$$t = 58°\ 56'\ 42'',$$
$$y = 76\ 43\ \ \ 6\ ,$$
$$x - t = 13\ 14\ 15\ ,$$
$$y - t = 17\ 46\ 26\ ,$$
$$z - t = 34\ 14\ 17\ ,$$
$$y - x = \ \ 4\ 32\ \ \ 7\ ,$$
$$z - y = 16\ 27\ 54\ .$$

In order to avoid large numbers in the calculation, let us, as in the previous example, assume from the observations approximate values of the angles and designate by t', y', etc., the corrections to be applied to those approximate values. Thus if we place

$$t = 58°\ 56'\ 42'' + t',$$
$$x = 72\ 10\ 57 + x',$$
$$y = 76\ 43\ \ \ 6 + y',$$
$$z = 93\ 10\ 58 + z'.$$

Then, by substituting these in the observation equations above, we get the simpler forms

$$t' = 0,$$
$$y' = 0,$$
$$x' - t' = 0,$$
$$y' - t' = 2'',$$
$$z' - t' = 1,$$
$$y' - x' = -2,$$
$$z' - y' = 2,$$

in which the numerical terms include seconds only. From these we form the normal equations (Arts. 33, 34)

$$4t' - x' - y' - z' = -3'',$$
$$-t' + 2x' - y' \quad\quad = 2,$$
$$-t' - x' + 4y' - z' = -2,$$
$$-t' \quad\quad - y' + 2z' = 3.$$

and by their solution find
$$t' = -0''\cdot 1, \quad x' = 1\cdot 0, \quad y' = 0\cdot 1, \quad \text{and} \quad z' = 1\cdot 5;$$
hence the most probable values of the direction angles are
$$t = 58° \ 56' \ 41''\cdot 9 = AOB,$$
$$x = 72 \ 10 \ 58 \ \cdot 0 = AOC,$$
$$y = 76 \ 43 \ \ 6 \ \cdot 9 = AOD,$$
$$z = 93 \ 10 \ 59 \ \cdot 5 = AOE,$$
from which by simple subtraction we may find any required angle as BOC, COD, etc. (Strictly speaking, this problem is a case of conditioned observations, Chap. IV. By taking, however, a limited number of unknown quantities, the idea of condition need not enter the work until the final operation of deducing any angle from the direction angles by subtraction; then, of course, the whole must equal the sum of its parts.)

Problems. 1. To determine the elevations of two points A and B above a datum O, measurements were made which give, A above $O = 12\cdot 3$ feet, B above $O = 27$ feet, and A below $B = 14\cdot 1$ feet. What are the most probable elevations of A and B? *Ans.* $B = 26\cdot 8$ feet, etc.

2. In the preceding example of the adjustment of angles, what are the best values if the sixth and seventh measurements were not taken?
Ans. $AOC = 72° \ 10' \ 57''$, etc.

Observations of Unequal Weight.

36. Suppose that four observations give each time the equation
$$a_1 s + b_1 t = M_1,$$
and that five other observations give each time
$$a_2 s + b_2 t = M_2.$$
The most probable values of s and t will be determined by writing the first equation four times, the second five times, and then by Art. 34 forming and solving the normal

equations. Considering for convenience only the unknown quantity s, we may write the equations

$$a_1 s + N_1 = 0, \text{ 4 times repeated,}$$
$$a_2 s + N_2 = 0, \text{ 5 times repeated.}$$

Forming from these the normal equations for s (Art. 33), we have

$$4 a_1 (a_1 s + N_1) + 5 a_2 (a_2 s + N_2) = 0.$$

If, instead of writing the observation equations 4 and 5 times respectively, we should multiply the first by $\sqrt{4}$ and the second by $\sqrt{5}$, giving

$$\sqrt{4}\, a_1 s + \sqrt{4}\, N_1 = 0,$$
$$\sqrt{5}\, a_2 s + \sqrt{5}\, N_2 = 0,$$

and then considering each of these as representing a single observation, form the normal equation for s by multiplying the first by $\sqrt{4}\, a_1$, the second by $\sqrt{5}\, a_2$, and adding the results, we would have

$$4 a_1 (a_1 s + N_1) + 5 a_2 (a_2 s + N_2) = 0,$$

which is the same as that found before.

To solve observation equations of unequal weight, we have then the following:—*Multiply each observation equation by the square root of its weight; then from these form and solve the normal equations as before.*

For illustration, let us suppose that four observations upon three unknown quantities have given the equations,

$$x + y + z = 5{\cdot}5 \quad \text{with the weight 3,}$$
$$x = 0{\cdot}053 \ldots\ldots\ldots\ldots\ldots 3,$$
$$y = 0{\cdot}003 \ldots\ldots\ldots\ldots\ldots 3,$$
$$z = -0{\cdot}043 \ldots\ldots\ldots\ldots\ldots 1,$$

that is, the last equation is the result of only one measurement, but each of the others of three. Multiplying each equation by the square root of its weight, we have

$$\sqrt{3}\, x + \sqrt{3}\, y + \sqrt{3}\, z = 5{\cdot}5\ \sqrt{3},$$
$$\sqrt{3}\, x = 0{\cdot}053\ \sqrt{3},$$
$$\sqrt{3}\, y = 0{\cdot}003\ \sqrt{3},$$
$$z = -0{\cdot}043.$$

UPON SEVERAL QUANTITIES. 55

Multiplying each equation in which x occurs by its coefficient, viz. $\sqrt{3}$, we form by addition of the results the normal equation for x; thus we have

$$6x + 3y + 3z = 16{\cdot}659,$$
$$3x + 6y + 3z = 16{\cdot}509,$$
$$3x + 3y + 4z = 16{\cdot}457,$$

which being solved give

$$x = 0{\cdot}9675, \quad y = 0{\cdot}9175, \quad z = 2{\cdot}7005,$$

which is the best system of values obtainable from the observations.

37. To adjust observations of unequal weight we have then, to write the observation equations, multiply each by the square root of its weight and from the reduced equations derive and solve the normal equations. The following examples will render the whole process clear.

I. *Level Lines.* The nine observations of Example 1, Art. 35, upon the elevations of the points S, T, etc. are of unequal weight. The least trustworthy is No. 9, because it is not known that mean tide at Baltimore is the same as the mean surface of the ocean, and we call its weight 1. Nos. 3 to 8 inclusive are ordinary railroad levels and may with reference to No. 8 be given a weight of 4. Nos. 1 and 2, being the result of carefully conducted government and canal levels extending over many years, are the most reliable of all, and we give them a weight of 25. The observation equations are the same as before; multiplying each by the square root of its weight, we have

$$5s = 2865{\cdot}40,$$
$$5s - 5t = 13{\cdot}00,$$
$$2t = 1150{\cdot}54,$$
$$2u - 2t = 334{\cdot}66,$$
$$2x - 2u = 7{\cdot}60,$$
$$2x - 2t = 340{\cdot}56,$$
$$2x - 2y = 850{\cdot}00,$$
$$2y = 639{\cdot}82,$$
$$y = 319{\cdot}75.$$

From these we form the normal equations by the usual method,

$$50s - 25t \qquad\qquad = 14262.00,$$
$$-25s + 37t - 4u - 4x \qquad = 1015.64,$$
$$-4t + 8u - 4x \qquad = 654.12,$$
$$-4t - 4u + 12x - 4y = 2396.32,$$
$$-4x + 9y = -100.61.$$

Before solution these may be simplified by dividing the first by 25, and the third and fourth by 4. The resulting elevations are

$$s = 572.98,$$
$$t = 575.48,$$
$$u = 742.36,$$
$$x = 745.72,$$
$$y = 320.25.$$

A comparison of these values with those deduced in Art. 35 will be interesting to the reader, as showing the influence of the weights and the closer agreement with Mr GARDNER's values. (See also Art. 40.)

II. *Adjustment of Angles.* In order to determine the angles AOB and BOC, Fig. 6, the instrument was set with its zero point in the direction OM and the following observations were taken.

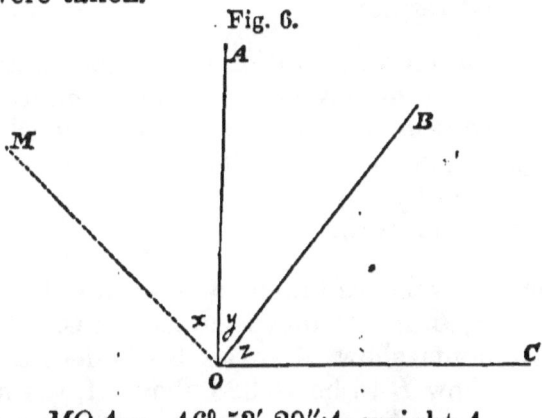

Fig. 6.

$$MOA = 46°\ 53'\ 29''.4, \text{ weight } 4,$$
$$MOB = 83\ 14\ 36\ .3, \ \dots\dots\ 16,$$
$$MOC = 135\ 27\ 11\ .7, \ \dots\dots\ 9;$$

the first being the mean of 4 readings, the second of 16, and the third of 9. What are the most probable values of MOA, AOB and BOC?

Let x, y and z represent the required angles, and to avoid using large numbers place

$$x = 46° 53' 30'' + x',$$
$$y = 36\ 21\ \ 7 + y',$$
$$z = 52\ 12\ 35 + z';$$

then the observation equations are

$$x' = -0''\cdot 6 \text{ with weight } 4,$$
$$y' + x' = -0\cdot 7 \ldots\ldots\ldots\ldots 16,$$
$$z' + y' + x' = -0\cdot 3 \ldots\ldots\ldots\ldots 9.$$

Applying the weights, forming and solving the normal equations, we find

$$x' = -0''\cdot 6,\ \ y' = 0''\cdot 24 \text{ and } z' = -0''\cdot 54,$$

and hence

$$x = 46° 53'\ 29''\cdot 4 = MOA,$$
$$y = 36\ 21\ \ \ 7\cdot 24 = AOB,$$
$$z = 52\ 12\ \ 34\cdot 46 = BOC,$$

are the adjusted results.

In the formation and solution of normal equations involving large numbers of observations, the computer will derive much assistance from the notation of GAUSS, which we have given in Part II., Arts. 37 and 58. In all cases the solution should be effected by independent methods, in order to check the accuracy of the work.

Problem. Levels are taken to determine the elevation of three points A, B and C above the datum O. The mean of five measurements show A to be 3·426 feet above O, the mean of nine show B to be 10·328 above A, the mean of four give C 2·471 above B: and lastly, one measurement gives B 13·762 feet above O. What are the most probable elevations of the three points?   *Ans.* $A = 3\cdot 4283$, etc.

Probable Errors and Weights.

38. The preceding methods are sufficient for the adjustment of any common indirect observations. We now come to the question, What degree of confidence can we place in the values of the quantities deduced from such measurements, or, in other words, to what degree of precision have we attained? This will be shown by the probable errors of those quantities (Art. 16).

Let n be the number of the measurements, having the relative weights g_1, g_2, g_3, etc.: the number of observation equations will also be n. Let q denote the number of unknown quantities s, t, u, etc. whose values are to be determined, and G_s, G_t, etc. their respective weights. When the most probable values of s, t, u, etc. have been found by the solution of the normal equations, let them be substituted in the observation equations (*not* in those equations after multiplication by the square roots of their weights); they will not reduce these equations to zero, but leave the residuals v_1, v_2, v_3, etc.

Let $\Sigma g v^2$ denote the sum $g_1 v_1^2 + g_2 v_2^2 + g_3 v_3^2 +$ etc., that is, the quantity formed by multiplying the square of each residual by the weight of the corresponding equation, and taking the sum of the products. Then, as proved in Part II., the probable error of an observation *of the weight unity* is

$$(87) \qquad r = 0.6745 \sqrt{\frac{\Sigma g v^2}{n - q}},$$

hence (Art. 30) the probable error of an observation of the weight g_1 is $\dfrac{r}{\sqrt{g_1}}$. The probable errors of the adopted values of s, t, u, etc. are also

$$(81) \qquad R_s = \frac{r}{\sqrt{G_s}}, \quad R_t = \frac{r}{\sqrt{G_t}}, \text{ etc.,}$$

to determine which we must compute r, and the weights G_s, G_t, etc. The weights g_1, g_2, etc. will be given by the conditions of the measurements. If, as in Art. 34, the observations are of equal weight, we have only to make all g's equal to 1. The weights G_s, G_t, of the determined quantities are as yet unknown: to find them GAUSS' method, as explained in the next article, may be employed.

39. As demonstrated in Part II., the weights of the values of the unknown quantities may be thus found. After each observation equation has been multiplied by the square root of its weight, and thus all reduced to the same unit of weight (Art. 36), let the normal equations be formed (Art. 33). These will be of the *form*

(80)
$$A_1 s + B_1 t + C_1 u + \text{etc.} = A,$$
$$A_2 s + B_2 t + C_2 u + \text{etc.} = B,$$
$$A_3 s + B_3 t + C_3 u + \text{etc.} = C,$$
$$\text{etc.}, \qquad \text{etc.},$$

the first being the normal equation for s, the second for t, the third for u, etc.; $A_1, A_2 \ldots B_1, B_2$, etc. being numerical coefficients of the unknown quantities, and A, B, C the absolute terms. The solution of these equations will give

(92)
$$s = a_1 A + a_2 B + a_3 C + \text{etc.},$$
$$t = \beta_1 A + \beta_2 B + \beta_3 C + \text{etc.},$$
$$u = \gamma_1 A + \gamma_2 B + \gamma_3 C + \text{etc.},$$
$$\text{etc.}, \qquad \text{etc.},$$

that is, each unknown quantity will be given in terms of the absolute terms A, B, C, etc. with the numerical coefficients $a_1, a_2, \beta_1, \beta_2$, etc. Then the weight of s is $\dfrac{1}{a_1}$, the weight of t is $\dfrac{1}{\beta_2}$, the weight of u is $\dfrac{1}{\gamma_3}$, and so on.

Stated in words the method is: *Place instead of the absolute terms in the normal equations the letters* A, B, C, *etc., and solve the equations. Then the weight of any unknown quantity as* s *is the reciprocal of the coefficient of the absolute term* A *in the value of* s, *the weight of* t *is the reciprocal of the coefficient of the absolute term* B *in the general value of* t, *etc.;* it being understood that A is the absolute term in the normal equation for s, B the absolute term in the normal equation for t, and so on.

Thus in Art. 35, Example 1, we have the normal equations

$$2s - t = A,$$
$$-s + 4t - u - x = B,$$
$$\text{etc.},$$

whose solution gives

$$s = \frac{32}{51}A + \frac{13}{51}B + \text{etc.},$$

$$t = \frac{13}{51}A + \frac{26}{51}B + \text{etc.},$$

etc.;

and hence the weights of the values of s, t, etc. are

$$\frac{51}{32}, \ \frac{51}{26}, \ \text{etc.}$$

Problem. Given the observation equations

$2s + t = 7$ with weight 3,
$s + 3t = 6$ 1,
$s - t = 2$ 4,

to find the values and weights of s and t.

Ans. $s = 3$ with weight 15·43,
$t = 1$ .... 14·53.

40. To illustrate in full the determination of weights and probable errors of indirectly observed quantities, let us take the example of Art. 37, in which the observation equations are

No. 1. $s = 573\cdot08$ with weight 25,
... 2. $t - s = 2\cdot60$ 25,
... 3. $t = 575\cdot27$ 4,
... 4. $u - t = 167\cdot33$ 4,
... 5. $x - u = 3\cdot80$ 4,
... 6. $x - t = 170\cdot28$ 4,
... 7. $x - y = 425\cdot00$ 4,
... 8. $y = 319\cdot91$ 4,
... 9. $y = 319\cdot75$ 1,

and the normal equations are

$$50s - 25t \qquad\qquad\qquad = 14262{\cdot}00 = A,$$
$$-25s + 37t - 4u - 4x = 1015{\cdot}64 = B,$$
$$-4t + 8u - 4x = 654{\cdot}12 = C,$$
$$-4t - 4u + 12x - 4y = 2396{\cdot}32 = D,$$
$$-4x + 9y = -100{\cdot}61 = E,$$

and it be required to find the probable error of each observation and the weights and probable errors of the values of t and x.

Placing the absolute terms equal to A, B, C, etc. and solving the equations by any algebraic method (the method of indeterminate multipliers is the shortest), we find

$$t = \frac{37A + 74B + 64C + 54D + 24E}{1341} = 575{\cdot}48,$$

$$x = \frac{36A + 72B + 171C + 270D + 120E}{1788} = 745{\cdot}72.$$

By the rule of Art. 39, the weight of t is the reciprocal of the coefficient of the absolute term B in the normal equation for t, or

$$G_t = \frac{1}{\frac{74}{1341}} = \frac{1341}{74} = 18{\cdot}12;$$

likewise the weight of x is the reciprocal of the coefficient of D, or

$$G_x = \frac{1788}{270} = 6{\cdot}62.$$

(Note. If the object be merely to determine the weight of t, it is evident that it is unnecessary to retain A, C, D and E in the algebraic work; they may be placed equal to zero. So in finding the weight of x, it is only necessary to retain D in the computation.)

We can now find the probable errors. The values of the quantities s, t, u, etc. as given by the normal equations are

$$s = 572{\cdot}98,$$
$$t = 575{\cdot}48 \text{ with weight } 18{\cdot}12,$$
$$u = 742{\cdot}36,$$
$$x = 745{\cdot}72 \text{ with weight } 6{\cdot}62,$$
$$y = 320{\cdot}25.$$

Inserting these in the observation equations, the remainders or residuals v_1, v_2, etc. are placed in the third column below, their squares in the fourth, and the product of each square by its corresponding weight in the fifth.

No.	g.	v.	v^2.	gv^2.
1.	25	0·10	0·010	0·250
2.	25	·11	·012	·300
3.	4	·20	·040	·160
4.	4	·44	·194	·776
5.	4	·43	·185	·720
6.	4	·02	·000	·002
7.	4	·48	·210	·840
8.	4	·34	·116	·464
9.	1	·50	·250	·250
			$\Sigma gv^2 =$	3·762

The sum of the products $g_1v_1^2$, $g_2v_2^2$, etc. is then 3·762. Since there are 9 observations and 5 unknown quantities, $n = 9$ and $q = 5$. Then (Art. 38) the probable error of an observation of weight unity, that is of No. 9, is

$$r = 0·6745 \sqrt{\frac{3·762}{4}} = 0·635 \text{ feet,}$$

and the probable error of observations 1 and 2 is

$$\frac{0·635}{5} = 0·107 \text{ feet,}$$

and of those from 3 to 8 inclusive is $\dfrac{0·635}{2} = 0·317$ feet. Also the probable errors of the above values of t and x are

$$R_t = \frac{0·635}{\sqrt{18·12}} = 0·153, \text{ and } R_x = \frac{0·635}{\sqrt{6·62}} = 0·248,$$

and hence we may write

$t =$ elevation of Cleveland datum $= 575·48 \pm 0·153$ feet,

$x =$ elevation of Pittsburg depot $= 745·72 \pm 0·248$.

It is then, as far as these observations show, an even wager that 575·48 feet does not differ from the true elevation of the Cleveland datum by 0·153, and also an even wager that 745·72 expresses the true height of the Pittsburg depot within 0·248 feet. The first is therefore much the more accurately determined.

Problem. Find the weights and probable errors of the values of $s, u,$ and y in the preceding example.

Ans. $s = 572\cdot98 \pm 0\cdot116$ feet, etc.

Other Applications.

By making $q = 1$ in the formulæ of Art. 38, they reduce to those of Arts. 24 and 30. Direct observations upon a single quantity are then only a particular case of indirect ones, and the methods of the present chapter are sufficient for their complete adjustment. Thus if there be only one unknown quantity z, and the measurements be made directly upon it, the observation equations are

$$z = M_1, \quad z = M_2, \quad z = M_3 \ldots z = M_n.$$

The normal equation formed from these will be

$$nz = M_1 + M_2 + M_3 + \ldots + M_n,$$

and this gives at once the rule of the average. By Art. 39 the weight of the value of z will be n, as in fact is implied in our definition of weight (Art. 27).

41. It is often the case that a quantity is measured in several parts, then of course its most probable value is the sum of the adjusted values of its parts. Thus, if we have measured by independent ways three parts of a base line, and find for them the values z_1, z_2, z_3, the value to be taken for the whole line is

$$Z = z_1 + z_2 + z_3.$$

If r_1, r_2 and r_3 are the probable errors of z_1, z_2 and z_3, the probable error of Z is

(100) $\qquad R = \sqrt{r_1^2 + r_2^2 + r_3^2}.$

In like manner, if a quantity Z is equal to any simple

function of the sum or difference of independently observed quantities, z_1, z_2, z_3, etc., that is, if
$$Z = z_1 \pm z_1 \pm z_2 \pm \text{etc.},$$
then having found the probable errors r_1, r_2, etc. of z_1, z_2, etc., the probable error of Z is given by the relation

(100) $\qquad R^2 = r_1^2 + r_2^2 + r_3^2 + \text{etc.}$

As an example illustrating this, let us take a case of levelling. In order to determine the difference of level between two points A and B, a level was set up halfway between them, and 20 readings taken on rods held at those points with the following results:

Rod at A.	Rod at B.
7 readings gave 7·229 feet,	3 readings gave 9·806,
8 .... 7·230 ...	12 .... 9·807,
5 .... 7·231 ...	5 .... 9·808.

What is the most probable difference of level between the two points and the probable error of the determination?

The general mean (Art. 27) of the readings at A is 7·2299, and of those at B, 9·8071 feet. Hence the difference of level is
$$9·8071 - 7·2299 = 2·5772 \text{ feet.}$$

To find its probable error, we have to find the probable errors of the two general means by Art. 30. For the mean 7·2299, we find
$$\Sigma gv^2 = 0·0000118, \; n = 3, \; G = 20, \text{ hence } R_1 = 0·00037,$$
and for the mean 9·8071, we have
$$\Sigma gv^2 = 0·0000078, \; n = 3, \; G = 20, \text{ hence } R_2 = 0·00031.$$

Then from the above principle the probable error of the difference 2·5772 is
$$\sqrt{0·00037^2 + 0·00031^2} = 0·00048.$$

Hence the adjusted results are

Reading at $A = 7·2299 \pm 0·00037$,
Reading at $B = 9·8071 \pm 0·00031$,
Diff. of level $= 2·5722 \pm 0·00048$.

Problem. The north declination of a star is
$$\delta = 19° 30' 14''\cdot 8$$
with a probable error of $0''\cdot 8$. The zenith distance of the same star is observed $\zeta = 21° 17' 20''\cdot 3$ with a probable error of $2''\cdot 3$. The latitude of the place of observation is $\phi = \delta + \zeta$. What is its probable error?

Ans. $\phi = 40° 47' 35''\cdot 1 \pm 2''\cdot 44$.

42. If we have two quantities Z and z_1 connected by the relation
$$Z = Az_1,$$
in which A is a constant, and if by measurement we find the value of z_1 and its probable error r, the probable error of Z is given by

(101) $$R = Ar_1.$$

Thus, the circumference of a circle is $3\cdot 1416$ times its diameter. If we measure the latter and find its value $z_1 = 1000 \pm 0\cdot 2$, the value and probable error of the former will be $Z = 3141\cdot 6 \pm 0\cdot 63$.

So in general if we have a quantity Z whose relation to the independent quantities z_1, z_2, z_3, etc. is given by
$$Z = Az_1 + Bz_2 + Cz_3 + \text{etc.},$$
and if we find the values of z_1, z_2, etc. and their probable errors r_1, r_2, r_3, etc. the probable error of Z is given by

(102) $$R^2 = A^2 r_1^2 + B^2 r_2^2 + C^2 r_3^2 + \text{etc.}$$

We have here given merely the statement of these important relations; the proof is presented in full in the second part of this work.

Miscellaneous Problems. 1. A chronometer is rated at a certain date and found to be $9^m\cdot 12^s\cdot 3$ fast with a probable error of $0^s\cdot 3$. Ten days afterwards it is again rated and found to be $9^m\cdot 21^s\cdot 4$ fast with the same probable error. What is the probable error of the mean daily rate?

The rate in the whole interval is
$$9^m\cdot 21^s\cdot 4 - 9^m\cdot 12^s\cdot 3 = 9^s\cdot 1,$$

with a probable error (Art. 41) of
$$\sqrt{0.3^2 + 0.3^2} = 0^s.42.$$
The mean daily rate is then $\frac{9^s.1}{10} = 0^s.91$ and its probable error (Art. 42) is
$$\frac{0^s.42}{10} = 0^s.042.$$
The clock gains then, daily, $9^s.1 \pm 0^s.042$.

2. Given the observation equations (all of equal weight)
$$\begin{aligned} 2x - y + z &= 3, \\ 3x + 3y - z &= 14, \\ 4x + y + 4z &= 21, \\ -5x + 2y + 3z &= 5, \end{aligned}$$
to find the best values of x, y, and z, and their probable errors.

Ans. $x = 1.916 \pm 0.026,$
$y = 3.551 \pm 0.052$, etc.

3. A block of cast iron weighing 100 lbs. rests upon a horizontal table also of cast iron. A horizontal force is applied to the block and it is observed that it begins to move when the force is 15.5 lbs. If the probable error in the determination of this force is 0.5 lbs., what is the probable error of the coefficient of friction μ?

Ans. $\mu = 0.155 \pm 0.005.$

4. The following levels were taken to determine the elevations of five points T, U, W, X and Y above the datum O:

T above $O = 115.52$,	X above $W = 632.25$,
U ...... $T = 60.12$,	X ...... $Y = 211.01$,
U ...... $O = 177.04$,	Y ...... $U = 596.12$,
W ...... $T = 234.12$,	Y ...... $W = 427.18$.
W ...... $U = 171.00$,	

What are the adjusted elevations?

Ans. $T = 115.61$, $U = 176.95$, etc.

5. An angle is measured by a theodolite giving the value 37° 16′ 13″ with a probable error of 5″, and also by a sextant giving 37° 16′ 10″ with a probable error of 3″. What is the adjusted value and its probable error?

Further examples illustrating the adjustment of indirect observations which arise in physical investigations are given in Chapter V.

CHAPTER IV.

CONDITIONED OBSERVATIONS.

43. ALL the observations thus far considered have been *independent*, so that an error committed in one measurement has had no connection whatever with those arising from the

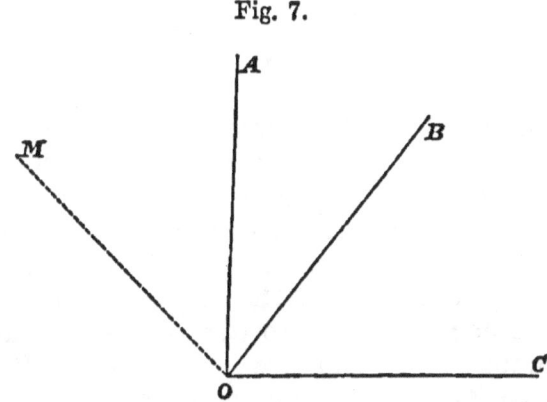

Fig. 7.

others. Thus in measuring the angles AOB and BOC, whether it be done directly or indirectly (Art. 21), each reading or observation has been entirely independent of those preceding and following, and a variation in the value of AOB does not necessarily require a corresponding one in BOC. But if a third observation be made upon AOC, the values of AOB, BOC and AOC are no longer independent, for since the whole must equal the sum of its parts, the relation $AOC = AOB + BOC$ exists, and it is impossible to suppose the value of BOC to vary without a corresponding variation in the value of AOC or AOB.

We have then a new class of equations, viz. *conditional equations*, which must be exactly satisfied by the values adopted for the unknown quantities, since they are the expression of axioms and rigorous laws. The number of

CONDITIONED OBSERVATIONS.

these equations is always less than the number of unknown quantities, for if they were as many in number, the values of the latter would be determined by their solution without the necessity of measurement. From these conditional equations and from the observation equations the values of the unknown quantities are to be found*.

Observations of equal weight.

44. The conditional equations being less than the number of unknown quantities may be satisfied in various ways. The observation equations cannot however be all exactly satisfied, and hence, as in the preceding chapter, the best system of values must be found. This is done (Arts. 16, 33) by making the sum of the squares of the residuals a minimum. Hence we have to determine the values of the unknown quantities in such a way *that they shall be the most probable values for the observation equations and shall at the same time exactly satisfy the conditional equations.*

Thus in the case of the last Article suppose that the measurements give the values

$$AOB = 36° \ 24' \ 30'',$$
$$BOC = 47 \ \ 52 \ \ 20,$$
$$AOC = 84 \ \ 17 \ \ 10.$$

If the value of AOC were exactly equal to the sum of the other two values, the quantities would need no correction. As they stand, however, the results are discordant and must be adjusted. Let the unknown values of AOB be x, of BOC, y, and of AOC, z, then we have the observation equations

$$x = 36° \ 24' \ 30'',$$
$$y = 47 \ \ 52 \ \ 20,$$
$$z = 84 \ \ 17 \ \ 10,$$

---

* In most books upon this subject, the term "equations of condition" is applied indiscriminately to both of these very distinct classes, and is a cause of some perplexity to the student. The excellent distinction of the Germans, *Beobachtungsgleichung* and *Bedingungsgleichung*, ought certainly to come into use.

and the rigorous conditional equation
$$x + y = z,$$
and we must determine the values of x, y and z so that they will exactly satisfy the latter and, at the same time, be the most probable values for the former. If for z, in the last observation equation, we place its value $x + y$ from the conditional equation we have

$$x = 36° \ 24' \ 30'',$$
$$y = 47 \ \ 52 \ \ 20,$$
$$x + y = 84 \ \ 17 \ \ 12,$$

or simply three observation equations, each one of which is now independent of the other, and which can be solved in the usual way (Arts. 33, 34). Multiplying the first and third by 1 and adding them, we have the normal equation for x, doing the same for the second and third we have the normal equation for y, or

$$2x + y = 120° \ 41' \ 40'',$$
$$x + 2y = 132 \ \ \ 9 \ \ 32.$$

The solution of these gives
$$x = 36° \ 24' \ 36\tfrac{2}{3}'', \quad y = 47° \ 52' \ 26\tfrac{2}{3}'',$$
and hence
$$z = x + y = 84° \ 17' \ 3\tfrac{1}{3}''.$$

The process here exhibited has already been illustrated in some of the examples of the preceding chapter, in which, instead of stating the conditional equations, we have simply written them as observation equations, omitting the additional unknown quantity. (See Art. 35. Example 2.) We may therefore give the following as the statement of a method for adjusting conditioned observations of equal weight.

1st. For every observation, whether direct or indirect, write an *observation equation* (Art. 32). Let n equal the number of such equations, and q the number of unknown quantities involved.

2nd. For each rigorous condition, write a *conditional equation* (Art. 43), and let p equal their number.

3rd. From the conditional equations find the values of p unknown quantities in terms of the others, and substitute them in the observation equations. There will then be n such equations containing $q - p$ unknown quantities, each of which will represent an *independent* observation.

4th. From these observation equations, form and solve the normal equations (Arts. 33, 34); the resulting values will be the most probable. Then the values of the remaining unknown quantities may be directly found from the conditional equations.

As a fuller illustration we choose the following. At the points A, B and C, there are measured the angles

$$s = 91°\ 27'\ 40'',$$
$$t = 43\ \ 52\ \ 50,$$
$$u = 44\ \ 39\ \ 50,$$
$$\overline{\text{sum} = 180\ \ \ 0\ \ 20,}$$

$$y = 20°\ 15'\ 10'',$$
$$z = 64\ \ 55\ \ 10,$$

Fig. 8.

subject to the geometrical conditions that at the point A

$$u + y = z,$$

and that in the triangle ABC

$$s + t + u = 180°.$$

To avoid the use of large numbers we adopt the method employed to some extent in the last chapter of assuming approximate values for the angles and regarding the corrections to be applied to those values as the unknown quantities; thus if we place

$$s = 91°\ 27' + s',$$
$$t = 43\ 52 + t',$$
$$\text{etc.}$$

and substitute the values in the above expressions, we have

$$s' = 40'', \qquad u' + y' = z' + 60,$$
$$t' = 50, \qquad s' + t' + u' = 120,$$
$$u' = 50,$$
$$y' = 10,$$
$$z' = 10,$$

in which the numbers represent seconds only. From the two conditional equations we take the values of any two unknown quantities in terms of the others, for example,

$$z' = u' + y' - 60, \quad s' = 120 - u' - t',$$

and substitute them in the observation equations, giving

$$t' + u' = 80'',$$
$$t' = 50,$$
$$u' = 50,$$
$$y' = 10,$$
$$y' + u' = 70.$$

From these we form the normal equations

$$2t' + u' \qquad = 130,$$
$$t' + 3u' + y' = 200,$$
$$u' + 2y' = 80,$$

whose solution gives

$$t' = 41''\cdot25, \quad u' = 47''\cdot5, \quad y' = 16''\cdot25.$$

Then from the conditional equations we have the remaining quantities

$$z' = 3''\cdot75, \quad s' = 31''\cdot25.$$

These are the corrections to be added to the approximate values assumed, that is, the number of seconds to be taken instead of the observed values. Hence the adjusted results are

$$s = 91° \ 27' \ 31''\cdot 25$$
$$t = 43 \ 52 \ 41 \ \cdot 25$$
$$u = 44 \ 39 \ 47 \ \cdot 50$$
$$\text{sum} = 180 \ \ 0 \ \ 0$$

$$u = 44° \ 39' \ 47''\cdot 50$$
$$y = 20 \ 15 \ 16 \ \cdot 25$$
$$z = 64 \ 55 \ 3 \ \cdot 75$$

which exactly satisfy the two geometrical conditions.

Problem. The three angles of a triangle were measured as follows:
$x = 98° \ 17' \ 22''$, $y = 70° \ 9' \ 56''$, and $z = 11° \ 32' \ 52''$. What are the adjusted values?

Ans. $x = 98° \ 17' \ 18'' \cdot 67$, etc.

45. The preceding method is general, and may be applied to any number of observations subject to any number of conditions. Although the simplest in theory, it is not always so in practice, particularly when the number of conditional equations is large. The process in most common use is "GAUSS' method of correlatives," which we will now proceed to state and exemplify, referring the reader to Part II. for the proof and the fuller algorithm of the method.

1st. Let the observations be adjusted by the methods of Chapter II., without reference to the conditional equations, and the results be called the measured values.

2nd. For each rigorous condition write a *conditional equation* (Art. 43). If the measured values exactly satisfy these they will need no correction; if not, a further adjustment is necessary.

3rd. Let s', t', u', etc. be corrections to be applied to these values to make them satisfy the conditional equations. Let the conditional equations expressed in terms of these corrections be of the form

(104) $\quad \begin{aligned} \alpha_1 s' + \alpha_2 t' + \alpha_3 u' + \text{etc.} &= N', \\ \beta_1 s' + \beta_2 t' + \beta_3 u' + \text{etc.} &= N'', \\ \gamma_1 s' + \gamma_2 t' + \gamma_3 u' + \text{etc.} &= N''', \end{aligned}$

in which α, β, γ, etc. and N denote known constants.

74   CONDITIONED OBSERVATIONS.

4th. For each one of the unknown quantities s', t', u', etc., write an *equation of correlative* containing as many new auxiliary unknown quantities as there are conditional equations. Let them be of the form

(110)
$$\alpha_1 K_1 + \beta_1 K_2 + \gamma_1 K_3 + \text{etc.} = s',$$
$$\alpha_2 K_1 + \beta_2 K_2 + \gamma_2 K_3 + \text{etc.} = t',$$
$$\alpha_3 K_1 + \beta_3 K_2 + \gamma_3 K_3 + \text{etc.} = u'$$
$$\text{etc.} \qquad \text{etc.}$$

in which K_1, K_2, K_3, etc. are the new auxiliary unknown quantities, K_1 having the same coefficients as in the first conditional equation, K_2 the same as in the second, etc.; so that the first vertical row of coefficients corresponds to the first horizontal row in the conditional equations. There will be as many equations of correlative as there are unknown quantities s', t', u', etc.

5th. From the equations of correlative let the *normal equations* (Art. 33) be formed, thus

(109)
$$(\alpha_1^2 + \alpha_2^2 + \text{etc.}) K_1 + (\alpha_1 \beta_1 + \alpha_2 \beta_2 + \text{etc.}) K_2 + \text{etc.}$$
$$= \alpha_1 s' + \alpha_2 t' + \text{etc.} = N',$$
$$(\alpha_1 \beta_1 + \alpha_2 \beta_2 + \text{etc.}) K_1 + (\beta_1^2 + \beta_2^2 + \text{etc.}) K_2 + \text{etc.}$$
$$= \beta_1 s' + \beta_2 t' + \text{etc.} = N'',$$

which will be as many in number as there are conditional equations, and whose absolute terms will be N', N'', etc. as given by the conditional equations. Solving these we find the values of K_1, K_2, etc.

6th. Substituting these values of K_1, K_2, etc. in the equations of correlative, we find the values of s', t', u', etc. which will exactly satisfy the conditional equations, and be the most probable corrections to the observed quantities.

This process will be better understood, and its simplicity over that of the preceding Article be seen, by a consideration of some practical examples. We take first that of Fig. 9, in which there are only two equations of condition.

CONDITIONED OBSERVATIONS.

At the points A, B and C there are measured five angles, each measurement having the same weight,

$$s = 91°\ 27'\ 40''$$
$$t = 43\ 52\ 50$$
$$u = 44\ 39\ 50$$
$$\overline{\text{sum} = 180\ 0\ 20}$$

$$u = 44°\ 39'\ 50''$$
$$y = 20\ 15\ 10$$
$$z = 64\ 55\ 10$$

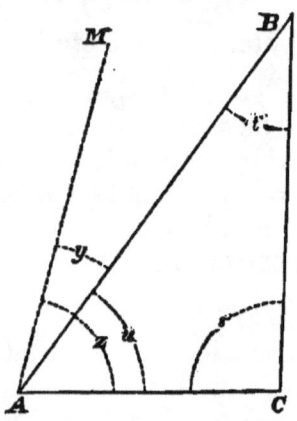

Fig. 9.

The rigorous geometrical conditions are

$$u + y = z,$$
$$s + t + u = 180°,$$

and since these are not satisfied by the measured angles, an adjustment is necessary. We assume the measured values as approximate, and take s', t', u', etc. as corrections to be applied to those values, then the conditional equations are

$$u' + y' - z' = 10'' \quad\dotfill(1),$$
$$s' + t' + u' = -20 \quad\dotfill(2).$$

We next assume two unknown auxiliary quantities K_1 and K_2, and for each unknown quantity write an equation of correlative, thus

$$K_2 = s' \quad\dotfill(3),$$
$$K_2 = t' \quad\dotfill(4),$$
$$K_1 + K_2 = u' \quad\dotfill(5),$$
$$K_1 = y' \quad\dotfill(6),$$
$$-K_1 = z' \quad\dotfill(7),$$

the coefficients of K_1 and K_2 in (3) being the same as those of s' in (1) and (2), viz. 0 and 1: in (5) the same as those of u' in (1) and (2), viz. 1 and 1: in (7) the same as those of z' in (1) and (2), viz. -1 and 0. From these we form the normal equations (Art. 33) for K_1 and K_2; viz.

$$3K_1 + K_2 = u' + y' - z' = 10,$$
$$K_1 + 3K_2 = s' + t' + u' = -20;$$

from which we have

$$K_1 = 6''\cdot 25 \text{ and } K_2 = -8''\cdot 75,$$

Substituting these values in (3), (4), (5), etc. we have

$s' = -8''\cdot 75$, $t' = -8\cdot 75$, $u' = -2\cdot 50$, $y' = 6\cdot 25$, $z' = -6\cdot 25$.

Applying these corrections to the measured angles, they become

$s' = 91°27'31''\cdot 25$
$t' = 435241\cdot 25$
$u' = 443947\cdot 50$
sum $= 180°00\cdot$

$u' = 44°39'47''\cdot 50$
$y' = 201516\cdot 25$
$z' = 64553\cdot 75$

or the same as deduced from the longer process of the preceding article.

46. As a second example we choose a case occurring in the common practice of every engineer, viz. the

Adjustment of the angles of a Quadrilateral. In order to determine the distances WZ, ZY, etc. (Fig. 10), the base line

Fig. 10.

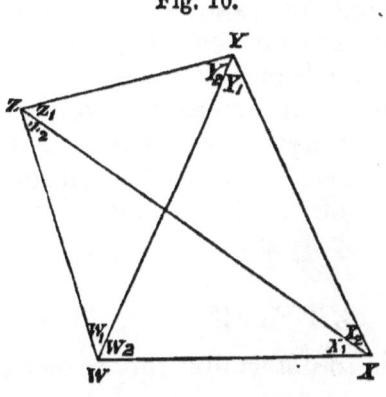

CONDITIONED OBSERVATIONS.

WX was measured and nine angles of the quadrilateral observed as follows (X denoting the large angle at the corner X, and X_1 and X_2 the small ones at the same corner, as shown in the sketch):

$$W = 106° \ 7' \ 30''$$
$$X_1 = \ 36 \ 34 \ 21$$
$$Z_2 = \ 37 \ 18 \ 12$$
$$\text{sum} = 180 \ \ 0 \ \ 3$$

$$X = \ 66° \ 34' \ 9''$$
$$Y_1 = \ 49 \ 17 \ 23$$
$$W_2 = \ 64 \ \ 8 \ 34$$
$$\text{sum} = 180 \ \ 0 \ \ 6$$

$$Z = \ 84 \ \ 7 \ 18$$
$$W_1 = \ 41 \ 58 \ 47$$
$$Y_2 = \ 53 \ 53 \ 50$$
$$\text{sum} = 179 \ 59 \ 55$$

In each of the four triangles which make up the figure, the sum of the measured angles differs from 180°. It will also be seen that at the corner W, the sum of the two small angles does not equal the large one, thus

$$W_1 = \ 41° \ 58' \ 47''$$
$$W_2 = \ 64 \ \ 8 \ 34$$
$$\text{sum} = 106 \ \ 7 \ 21 \quad \text{while } W = 106° \ 7' \ 30'';$$

and the sum of the four quadrilateral angles is not 360°. The problem before us is, to adjust these angles so that in every triangle the sum of the three angles shall be exactly 180°, so that at every corner the large angle shall equal the sum of the two small ones, so that the sum of the four quadrilateral angles shall be exactly 360°, and further, so that the adjusted values shall differ from the measured values by the least possible amounts.

We assume the above measured values as approximate ones, and take as unknown quantities the corrections to be applied to those values. Thus, at the corner W we designate the measured value of the large angle by W, the correction to be applied to it by w, and the required adjusted value by XWZ, so that

$$XWZ = W + w, \quad ZWY = W_1 + w_1, \quad WZX = Z_2 + z_2, \text{ etc.},$$

and the problem is to determine the corrections w, w_1, z_2, etc.

In order to avoid too many equations we select any corner as W, and take the three triangles WXZ, ZWY, and XYW which meet at that point as the three triangles for correction: the angles of the fourth triangle XYZ will be found by simple addition and subtraction as soon as those of the other three are determined.

The conditional equations for these three triangles are next to be stated. In the triangle WXZ the corrections to be applied to the measured angles are w, x_1, and z_2; and since the measured angles add up to $180° 0' 3''$, the algebraic sum of the three corrections must equal $-3''$; and hence the first conditional equation,

$$w + x_1 + z_2 = -3 \quad \text{.....................(1)}.$$

Next, in the triangle ZWY the corrections are z, w_1, and y_2, and they must exactly balance the discrepancy between $180°$ and the sum of Z, W_1, and Y_2, or

$$z + w_1 + y_2 = 5 \quad \text{.....................(2)}.$$

Likewise in the triangle XYW, we must have

$$x + y_1 + w_2 = -6 \quad \text{.....................(3)}.$$

Also at the corner W the corrections w, w_1 and w_2 must exactly balance the discrepancy of $9''$ between w and the sum of w_1 and w_2, or we must have

$$w_1 + w_2 - w = 9 \quad \text{.....................(4)}.$$

These are the four geometrical conditions which the corrections must exactly satisfy. From these four equations we must find the values of the nine unknown quantities in such a way that they shall exactly satisfy the conditional equation, and be the most probable system of corrections. This may be done either by the method of Art. 44, or by that of Art. 45. By the former we could write nine observation equations (viz. $w = 0$, $x_1 = 0$, $z_2 = 0$, etc.), then from the five conditional equations find the values of five unknown quantities in terms of the other four, and substitute them in the observation equations, then from those nine equations deduce the four normal equations, and by their solution find the required corrections. We shall however follow the "method of correlatives," since it is by far the easier and shorter.

CONDITIONED OBSERVATIONS.

Our conditional equations are then the following:

$$w + x_1 + z_2 = -3 \quad \ldots \ldots \ldots \ldots \ldots (1),$$
$$z + w_1 + y_2 = 5 \quad \ldots \ldots \ldots \ldots \ldots (2),$$
$$x + y_1 + w_2 = -6 \quad \ldots \ldots \ldots \ldots \ldots (3),$$
$$w_1 + w_2 - w = 9 \quad \ldots \ldots \ldots \ldots \ldots (4).$$

We assume four auxiliary unknown quantities $K_1, K_2, K_3,$ and K_4, and for each unknown quantity write an equation of correlative, thus

$$+ K_1 \qquad\qquad - K_4 = w \quad \ldots \ldots \ldots \ldots (5),$$
$$+ K_2 \qquad + K_4 = w_1 \quad \ldots \ldots \ldots \ldots (6),$$
$$+ K_3 + K_4 = w_2,$$
$$+ K_3 \qquad = x,$$
$$+ K_1 \qquad\qquad = x_1,$$
$$+ K_3 \qquad = y_1,$$
$$+ K_2 \qquad = y_2,$$
$$+ K_2 \qquad = z,$$
$$+ K_1 \qquad\qquad = z_2;$$

the coefficients of K_1 being the coefficients of the corresponding unknown quantities in equation (1), the coefficients of K_3 being those of the unknown quantities in equation (3), and so on. From these equations we form the normal equations for $K_1, K_2,$ etc., viz.

$$3K_1 \qquad\qquad - K_4 = -3 = w + x_1 + z_2,$$
$$3K_2 \qquad + K_4 = +5, = \text{etc.}$$
$$3K_3 + K_4 = -6,$$
$$-K_1 + K_2 + K_3 + 3K_4 = +9,$$

whose solution gives

$$K_1 = 0{\cdot}389, \quad K_2 = 0{\cdot}278, \quad K_3 = -3{\cdot}389, \quad K_4 = 4{\cdot}167.$$

Substituting these in the equations of correlative (5), (6), etc. we have

CONDITIONED OBSERVATIONS.

$$w = K_1 - K_4 = -3''{\cdot}78,$$
$$w_1 = K_2 + K_4 = 4{\cdot}44,$$
$$w_2 = K_3 + K_4 = 0{\cdot}78,$$
$$x = K_3 = -3{\cdot}39,$$
$$x_1 = K_1 = 8{\cdot}39,$$
$$y_1 = K_3 = -3{\cdot}39,$$
$$y_2 = K_2 = 0{\cdot}28,$$
$$z = K_2 = 0{\cdot}28,$$
$$z_2 = K_1 = 0{\cdot}39.$$

Applying these corrections to the measured angles, they become

$W + w_1 = 106°\ 7'\ 26''{\cdot}22$ $\quad X + x = 66°\ 34'\ 5''{\cdot}61$
$X_1 + x_1 = \ \ 36\ 34\ 21\ {\cdot}39$ $\quad Y_1 + y_1 = \ \ 49\ 17\ 19\ {\cdot}61$
$Z_2 + z_2 = \ \ 37\ 18\ 12\ {\cdot}39$ $\quad W_2 + w_2 = \ \ 64\ \ 8\ 34\ {\cdot}78$
$\text{sum} = 180\ \ 0\ \ 0\ {\cdot}0$ $\quad\quad \text{sum} = 180\ \ 0\ \ 0{\cdot}0$

$Z + z = \ \ 84°\ 7'\ 18''{\cdot}28$
$W_1 + w_1 = \ \ 41\ 58\ 51\ {\cdot}44$ $\quad W_1 + w_1 = 41°\ 58'\ 51''{\cdot}44$
$Y_2 + y_2 = \ \ 53\ 53\ 50\ {\cdot}28$ $\quad W_2 + w_2 = \ \ 64\ \ 8\ 34\ {\cdot}78$
$\text{sum} = 180\ \ 0\ \ 0\ {\cdot}0$ $\quad \text{sum} = 106\ \ 7\ 26\ {\cdot}22$
$\quad\quad\quad\quad\quad\quad\quad\quad\quad\quad = W + w.$

The three triangles WXZ, ZWY, and XYW are hence exactly adjusted, as are also the angles at the corner W. The remaining angles are obtained by simple addition and subtraction of those already found, thus

$Y + y = (Y_1 + y_1) + (Y_2 + y_2) = 103°\ 11'\ 9''{\cdot}89$
$Z_1 + z_1 = (Z + z) - (Z_2 + z_2) = \ \ 46\ 49\ 5\ {\cdot}89$
$X_2 + x_2 = (X + x) - (X_1 + x_1) = \ \ 29\ 59\ 44\ {\cdot}22$
$\quad\quad\quad\quad\quad\quad\quad\quad \text{sum} = 180\ \ 0\ \ 0\ {\cdot}0$

and the angles of the quadrilateral also add up exactly to $360°$.

. The above adjustment is sufficient for cases arising in common engineering practice. In extensive triangulations, however, where the sides are many miles in length, a fifth conditional equation is necessary. This arises from the fact that we have not considered the relation existing between

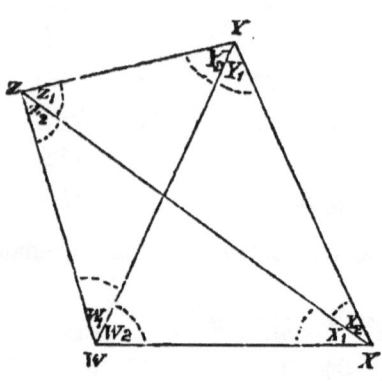

Fig. 10.

the sides and angles, so that if with the above adjusted angles we were to compute from the base WX the side ZY, through the triangles WZX and ZYX, and again through the triangles WYZ and WZY, the two results would not exactly agree. In such triangulations, therefore, as are made in our Coast Survey, besides the conditional equations between the angles, others called *side equations* are introduced. We proceed to illustrate the development and application of this relation between the sides to the case of the above quadrilateral.

Designating the adjusted angles by the same notation as before, we have the condition that in each of the three triangles WZX, WZY and WYZ, when adjusted, the sides must be proportional to the sines of their opposite angles, or

$$\frac{WX}{WZ} = \frac{\sin WZX}{\sin WXZ}, \quad \frac{WZ}{WY} = \frac{\sin WYZ}{\sin WZY}, \quad \frac{WY}{WX} = \frac{\sin WXY}{\sin WYX};$$

multiplying these equations together, member by member, we have

$$\frac{\sin WZX \sin WYZ \sin WXY}{\sin WXZ \sin WZY \sin WYX} = 1,$$

which is the fifth equation of condition. To adapt it to numerical computation, we write it

$$\sin WXY \sin WYZ \sin WZX = \sin WXZ \sin WYZ \sin WZY.$$

M.

Now WXY, WYZ, etc. are the adjusted values of the angles, or
$$WXY = X + x, \quad WYZ = Y_2 + y_2, \text{ etc.};$$
hence the equation is
$$\sin(X + x) \sin(Y_2 + y_2) \sin(Z_2 + z_2)$$
$$= \sin(X_1 + x_1) \sin(Y_1 + y_1) \sin(Z + z),$$
in which x, y_2, &c. are the unknown corrections. Applying logarithms, the equation becomes
$$\log \sin(X + x) + \log \sin(Y_2 + y_2) + \log \sin(Z_2 + z_2)$$
$$= \log \sin(X_1 + x_1) + \log \sin(Y_1 + y_1) + \log \sin(Z + z).$$

Now the corrections x, y_2, z_2, &c. are very small, each being only a few seconds of arc. Therefore very nearly
$$\log \sin(X + x) = \log \sin X + x \log. \text{ diff. } 1'',$$
$$\log \sin(Y_2 + y_2) = \log \sin Y_2 + y_2 \log. \text{ diff. } 1'', \text{ etc.,}$$
in which log. diff. $1''$ denotes the tabular logarithmic difference for $1''$ corresponding to the measured values X, Y_2, etc. Hence, inserting these values and transposing, our equation becomes

$x_1 \text{diff.}1'' + y_2 \text{diff.}1'' + z_2 \text{diff.}1'' - x_1 \text{diff.}1'' - y_1 \text{diff.}1'' - z_2 \text{diff.}1''$
$= \log \sin X_1 + \log \sin Y_1 + \log \sin Z - \log \sin X - \log \sin Y_2$
$\qquad - \log \sin Z_2.$

The logarithmic sines of the measured angles and their tabular differences may then be taken from a table and arranged as below:

Angle	log. sin	diff. 1″	Angle	log. sin	diff. 1″
X_1	9·77512924	0·00000284	X	9·96262539	0·00000091
Y_1	9·87967913	0·00000181	Y_2	9·90739060	0·00000154
Z	9·99771028	0·00000021	Z_2	9·78249744	0·00000277
sum = 29·65251865			sum = 29·65251343		

CONDITIONED OBSERVATIONS. 83

Substituting these values the equation is
$$0{\cdot}00000091x + 0{\cdot}00000154y_2 + \text{etc.}$$
$$= 29{\cdot}65251865 - 29{\cdot}65251343;$$
or after multiplying by 10000000 to avoid decimals,
$$91x + 154y_2 + 277z_2 - 284x_1 - 181y_1 - 21z = 522 \ldots\ldots(5)$$
which is the conditional equation, that any side computed from the measured base shall always be of the same length, whatever set of triangles be employed.

The four conditions between the angles are the same as before; and we have to determine the nine unknown corrections so that they shall be the best values, and at the same time exactly satisfy the five conditions,

$$w + x_1 + z_2 = -3 \ \ldots\ldots\ldots\ldots\ldots\ldots(1),$$
$$z + w_1 + y_2 = 5 \ \ldots\ldots\ldots\ldots\ldots\ldots(2),$$
$$x + y_1 + w_2 = -6 \ \ldots\ldots\ldots\ldots\ldots\ldots(3),$$
$$w_1 + w_2 - w = 9 \ \ldots\ldots\ldots\ldots\ldots\ldots(4),$$
$$91x + 154y_2 + 277z_2 - 284x_1 - 181y_1 - 21z = 522\ldots(5).$$

This we perform as before by the method of correlatives. Assuming five auxiliary unknown quantities, K_1, K_2, etc., the equations of correlatives are

$$\begin{aligned}
+K_1   -K_4 &= w, \\
+K_2 +K_4 &= w_1, \\
 +K_3 +K_4 &= w_2, \\
 +K_3 +91K_5 &= x, \\
+K_1   -284K_5 &= x_1, \\
 +K_3 -181K_5 &= y_1, \\
+K_2   +154K_5 &= y_2, \\
+K_2   -21K_5 &= z, \\
+K_1   +277K_5 &= z_2;
\end{aligned}$$

the coefficients of K_1 being those of the corresponding unknown quantities in the conditional equation (1), the coefficients of K_5 being those of the corresponding unknown quantities in equation (5), etc. From these equations the normal equations for K_1, K_2, etc. are formed, viz.

$$3K_1 \qquad\qquad -K_4- \quad 7K_5 = -3,$$
$$\qquad 3K_2 \qquad +K_4+ \quad 133K_5 = 5,$$
$$\qquad\qquad 3K_3+K_4- \quad 90K_5 = -6,$$
$$-K_1+ \quad K_2+ \quad K_3+3K_4 \qquad\qquad = 9,$$
$$-7K_1+133K_2- 90K_3 \qquad +222584K_5 = 522.$$

Solving these equations by any algebraic method, we have

$K_1 = 0{\cdot}393,\ K_2 = 0{\cdot}238,\ K_3 = -3{\cdot}366,\ K_4 = 4{\cdot}174,\ K_5 = 0{\cdot}000855.$

Substituting these in the equations of correlative we get the values of the required corrections, viz.

$$w = K_1 - \quad K_4 = -3''{\cdot}78,$$
$$w_1 = K_2 + \quad K_4 = 4{\cdot}41,$$
$$w_2 = K_3 + \quad K_4 = 0{\cdot}81,$$
$$x = K_3 + 91K_5 = -3{\cdot}29,$$
$$x_1 = K_1 - 284K_5 = 0{\cdot}15,$$
$$y_1 = K_3 - 181K_5 = -3{\cdot}52,$$
$$y_2 = K_2 + 154K_5 = 0{\cdot}37,$$
$$z = K_2 - 21K_5 = 0{\cdot}22,$$
$$z_2 = K_1 + 277K_5 = 0{\cdot}63.$$

These values exactly satisfy the conditional equations (1), (2), (3), and (4), and almost satisfy (5). Applying them to the measured angles we have the adjusted values, which differ by only one or two-tenths of a second from those determined without the use of the side equation, but which conform equally well to the geometrical requirements of the angles, and which in delicate computations will give no discordant results in the lengths of the sides.

The student who verifies the above results will have acquired some idea of the processes involved in the adjustment of extensive triangulations, for the method here given is that employed by the computers of the U. S. Coast Survey

CONDITIONED OBSERVATIONS.

Office. As the number of triangles increases the number of conditional equations and unknown quantities rapidly multiplies, and their proper solution requires great skill and patience. In the *U. S. Coast Survey Report* for 1854 mention is made of the adjustment of a triangulation involving *sixty-five* conditional equations. In such cases, we ought to say, the work is reduced to a routine by means of the algorithm exhibited in Part II., Arts. 37 and 60, and the results are obtained independently by different computers. In fact, it would be well-nigh impossible to deal with such a large number of equations, without performing the numerical operations by a systematic routine, and employing frequent checks to test the accuracy of the work.

As an exercise in operating with side equations we give the following problem, which requires no angle equations and only one side equation, which will be similar in form to that of the preceding example. The answers to the problem may be seen in the *U. S. Coast Survey Report* for 1854, Appendix, p. 82.

Problem. In the quadrilateral $WXYZ$, Fig. 10, five angles were measured as follows:

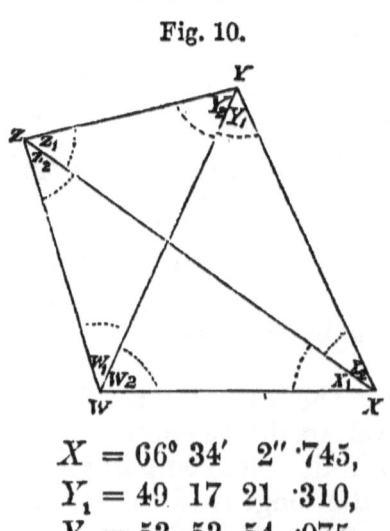

Fig. 10.

$X = 66° \; 34' \;\; 2'' \; \cdot 745,$
$Y_1 = 49 \;\; 17 \;\; 21 \;\; \cdot 310,$
$Y_2 = 53 \;\; 53 \;\; 54 \;\; \cdot 075,$
$Z_1 = 46 \;\; 49 \;\;\; 5 \;\; \cdot 167,$
$Z_2 = 37 \;\; 18 \;\; 11 \;\; \cdot 542.$

What are the adjusted values, so that in computing any side from the measured base *WX*, it shall always be of the same length whatever set of triangles be employed?

Observations of Unequal Weight.

47. A general method of adjusting conditioned observations of different weights follows directly from the processes of Arts. 37 and 44. It may be thus stated:

1st. For each of the n observations write an *observation equation* (Art. 32) in terms of the unknown quantities.

2nd. For each rigorous requirement write a *conditional equation* (Art. 43).

3rd. From the conditional equations find the values of p unknown quantities in terms of the others, and substitute them in the n observation equations, each of which will then be independent (Art. 44).

4th. Multiply each observation equation by the square root of its weight (Art. 37).

5th. Form and solve the *normal equations* (Arts. 33, 34), thus obtaining the most probable values of the unknown quantities.

For example, suppose that at the point O, Fig. 11, four angles are measured as follows:

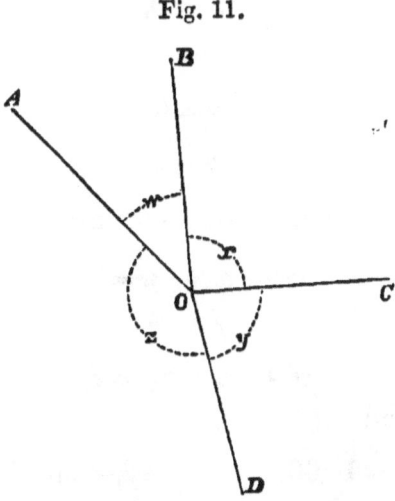

Fig. 11.

CONDITIONED OBSERVATIONS.

$$AOB = 40° \; 52' \; 37'' \quad \text{weight } 16$$
$$BOC = 92 \;\; 25 \;\; 41 \quad \text{''} \quad\quad 4$$
$$COD = 80 \;\;\; 6 \;\; 15 \quad \text{''} \quad\quad 3$$
$$\underline{DOA = 146 \;\; 35 \;\; 20 \quad \text{''} \quad\quad 1}$$
$$\text{sum} = 359 \;\; 59 \;\; 53$$

the angle AOB being the mean of 16 observations, BOC of 4, etc.

We assume the measured angles as approximate, and take w, x, y, and z as corrections to be applied to those values. Then the observation equations are

$$w = 0 \text{ with weight } 16,$$
$$x = 0 \quad \text{''} \quad\quad \text{''} \quad\quad 4,$$
$$y = 0 \quad \text{''} \quad\quad \text{''} \quad\quad 3,$$
$$z = 0 \quad \text{''} \quad\quad \text{''} \quad\quad 1,$$

and the rigorous conditional equation is

$$w + x + y + z = 7''.$$

Taking from this the value of z, inserting it in the observation equations, and multiplying each by the square root of its weight, we have

$$4w = 0,$$
$$2x = 0,$$
$$\sqrt{3}y = 0,$$
$$w + x + y = 0.$$

From these the normal equations are

$$17w + x + y = 7,$$
$$w + 5x + y = 7,$$
$$w + x + 4y = 7,$$

from which we find

$$w = 0''\cdot27, \quad x = 1''\cdot06, \quad y = 1''\cdot42, \text{ and hence } z = 4''\cdot25.$$

Adding these to the measured values, we have

$$AOB = 40° \; 52' \; 37'' \cdot 27$$
$$BOC = 92 \; 25 \; 42 \; \cdot 06$$
$$COD = 80 \; \; 6 \; 16 \; \cdot 42$$
$$DOA = 146 \; 35 \; 24 \; \cdot 25$$
$$\text{sum} = 360 \; \; 0 \; \; 0 \; \cdot 0$$

Problem. To determine the angles of the triangle ABC the following measurements were taken.

Fig. 12.

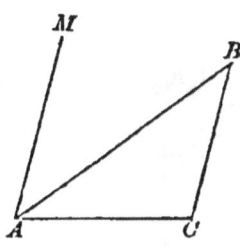

$ACB = 91° \; 27' \; 35''$ weight 4,
$CBA = 43 \; 52 \; 42$ „ 9,
$MAC = 64 \; 55 \; 5$ „ 2,
$MAB = 20 \; 15 \; 15$ „ 2.

What are the adjusted results?

Ans. $ACB = 91° \; 27' \; 33'' \cdot 71$,
$CBA = 43 \; 52 \; 41 \cdot 43$, etc.

48. As in the case of equal weighted observations, GAUSS' method of correlatives may be advantageously employed when the number of unknown quantities or conditional equations is greater than two or three. Although the process of Art. 47 is perfectly general and applicable to all cases, it is not in practice of so easy application as the method of correlatives, which is hence generally employed. The following is the statement of the process to be followed:

CONDITIONED OBSERVATIONS.

1st. Combine the observations by the methods of Chapters II. and III., and find their most probable values, and their weights.

2nd. Let s', t', u', etc. be the best system of corrections to be applied to those values to make them conform to the conditional equations, which expressed in terms of those corrections are of the form,

(104)
$$a_1 s' + a_2 t' + a_3 u' + \text{etc.} = N',$$
$$\beta_1 s' + \beta_2 t' + \beta_3 u' + \text{etc.} = N''.$$

3rd. Let g_s, g_t, etc. be the weights of the values to which s', t', etc. are corrections, and for each of these corrections write an *equation of correlative*, thus

(110)
$$\frac{1}{g_s}\left(a_1 K_1 + \beta_1 K_2 + \gamma_1 K_3 + \text{etc.}\right) = s',$$
$$\frac{1}{g_t}\left(a_2 K_1 + \beta_2 K_2 + \gamma_2 K_3 + \text{etc.}\right) = t',$$
$$\frac{1}{g_u}\left(a_3 K_1 + \beta_3 K_2 + \gamma_3 K_3 + \text{etc.}\right) = u', \text{ etc.,}$$

in which K_1, K_2, etc. are new auxiliary quantities, K_1 having the same coefficients as in the first conditional equations, K_2 the same as in the second, etc.

4th. From these form the *normal equations*, for K_1, K_2, etc. (not regarding $\dfrac{1}{g_s}$, $\dfrac{1}{g_t}$, etc., as coefficients of K_1, K_2, etc.). Then the second terms of these normal equations will be N', N'', etc., as in the given conditional equations.

5th. Solve the normal equations, and then substitute the resulting values of K_1, K_2, etc. in the correlative equations, thus finding the best system of corrections.

Although somewhat tedious to state, this method is of very rapid application. As a first example we choose a case involving but three conditional equations.

At the points A, B, C, and D, Fig. 13, there are measured the following seven angles.

CONDITIONED OBSERVATIONS.

Fig. 13.

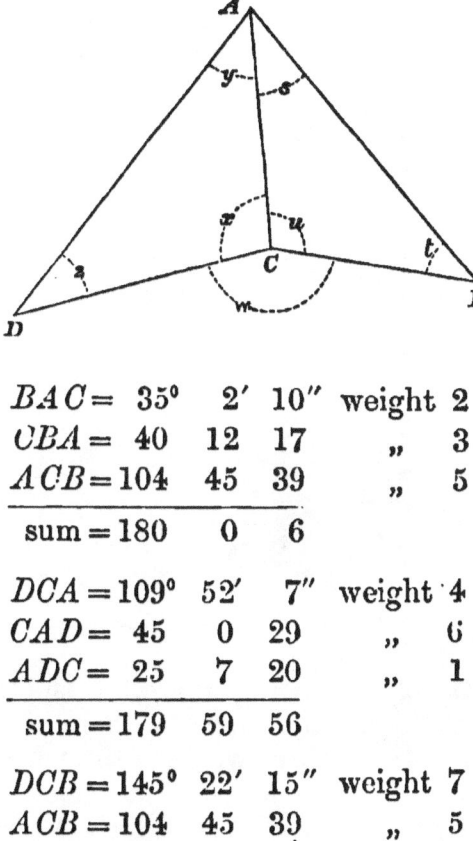

$BAC =$	35°	2'	10"	weight 2
$CBA =$	40	12	17	„ 3
$ACB =$	104	45	39	„ 5
sum =	180	0	6	
$DCA =$	109°	52'	7"	weight 4
$CAD =$	45	0	29	„ 6
$ADC =$	25	7	20	„ 1
sum =	179	59	56	
$DCB =$	145°	22'	15"	weight 7
$ACB =$	104	45	39	„ 5
$DCA =$	109	52	7	„ 4
sum =	360	0	1	

the angles of the triangle ABC being 6" greater than 180°, those of ACD being 4" less than 180°, and those around the point C being 1" greater than 360°.

We assume the measured angles as approximate, and take $s, t, u,$ etc. as corrections to be applied to $BAC, CBA,$ etc. (see Fig. 13). Then the conditional equations are

$$s + t + u = -6" \quad \text{...........................(1)},$$
$$x + y + z = 4 \quad \text{...........................(2)},$$
$$u + w + x = -1 \quad \text{...........................(3)}.$$

CONDITIONED OBSERVATIONS. 91

The weights of s, t, u, w, x, y and z are 2, 3, 5, 7, 4, 6 and 1, corresponding to the weights of the measured angles to which they are corrections. The equations of correlative are then

$$\tfrac{1}{2}(+K_1 \qquad\qquad) = s,$$

$$\tfrac{1}{3}(+K_1 \qquad\qquad) = t,$$

$$\tfrac{1}{5}(+K_1 \qquad + K_3) = u,$$

$$\tfrac{1}{7}(\qquad\qquad + K_3) = w,$$

$$\tfrac{1}{4}(\qquad + K_2 + K_3) = x,$$

$$\tfrac{1}{6}(\qquad + K_2 \qquad) = y,$$

$$\tfrac{1}{1}(\qquad + K_3) = z,$$

the coefficients of K_1 being those of the corresponding unknown quantities in equation (1), those of K_2 from those of (2) and so on. From these we form the normal equations

$$\left(\tfrac{1}{2}+\tfrac{1}{3}+\tfrac{1}{5}\right)K_1 + \tfrac{1}{5}K_3 \qquad = s+t+u = -6,$$

$$\left(\tfrac{1}{4}+\tfrac{1}{6}+1\right)K_2 + \tfrac{1}{4}K_3 \qquad = x+y+z = 4,$$

$$\tfrac{1}{5}K_1 + \tfrac{1}{4}K_2 + \left(\tfrac{1}{5}+\tfrac{1}{4}+\tfrac{1}{7}\right)K_3 = u+x+w = -1,$$

or after reduction

$$31K_1 \qquad\qquad + 6K_3 = -180,$$
$$34K_2 + 6K_3 = 96,$$
$$28K_1 + 35K_2 + 83K_3 = -140,$$

from which we have by solution

$$K_1 = -5\text{·}601, \quad K_2 = +3\text{·}01, \quad K_3 = -1\text{·}058.$$

Substituting these values in the correlative equations, we have

$s' = -2''\cdot 80, \quad t = -1\cdot 87, \quad u = -1\cdot 33, \quad w = -0\cdot 15,$
$x = 0\cdot 49, \quad y = 0\cdot 50, \quad z = 3\cdot 01.$

Adding these to the measured values we obtain

$BAC =$	35°	2′	7″·20	$DCA =$	109°	52′	7″·49
$CBA =$	40	12	15 ·13	$CAD =$	45	0	29 ·50
$ACB =$	104	45	37 ·67	$ADC =$	25	7	23 ·01
sum =	180	0	0 ·0	sum =	180	0	0 ·0

$DCB =$	145°	22′	14″·85
$ACB =$	104	45	37 ·67
$DCA =$	109	52	7 ·47
sum =	360	0	0 ·0

Which is the best system of adjusted angles?

Problems. 1. In a quadrilateral $WXYZ$ the four angles are measured as follows:

$W = 65° 11' 54''$ with weight 5,
$X = 66 24 15\cdot5 $ „ „ 10,
$Y = 87 2 24\cdot7 $ „ „ 12,
$Z = 141 21 20\cdot $ „ „ 1.

What are the adjusted angles?

Ans. $W = 65° \; 11' \; 54''\cdot 84$, etc.

2. In a *spherical* triangle XYZ the three angles were measured

$X = 93° \; 48' \; 15''\cdot 22$ with weight 30,
$Y = 51 \; 55 0 \cdot 18 $ „ „ 19,
$Z = 34 \; 16 \; 49 \cdot 72 $ „ „ 13.

The spherical excess is $4''\cdot 054$. What are the adjusted angles?

The conditional equation is
$$X + Y + Z = 180° \ 0' \ 4''{\cdot}054,$$
or taking x, y, and z as corrections, it is
$$x + y + z = -1''{\cdot}066.$$
The correlative equations then are
$$\frac{K_1}{30} = x, \quad \frac{K_1}{19} = y, \quad \text{and} \ \frac{K_1}{13} = z.$$
From which the normal equation is
$$\left(\frac{1}{30} + \frac{1}{19} + \frac{1}{13}\right) K_1 = -1''{\cdot}066, \ \text{or} \ K_1 = -6{\cdot}544;$$
and hence the corrections are
$$x = -0{\cdot}217, \quad y = -0{\cdot}347, \quad z = -0{\cdot}503,$$
which applied to the measured angles, will adjust them so as to agree with the geometrical condition for a spherical triangle.

3. To adjust the percentages of the following chemical analysis of pig iron, so that the sum shall be 100.

Carbon	2·30	per cent.—	weight of determination			= 1
Silicon	1·36	,,	,,	,,	,,	5
Phosphorus	1·02	,,	,,	,,	,,	5
Sulphur	0·61	,,	,,	,,	,,	5
Iron	94·93	,,	,,	,,	,,	$\frac{1}{10}$
Manganese	0·81	,,	,,	,,	,,	1

Ans. Sulphur = 0·594. Iron = 94·11 etc.

4. A base line AC is found by one measurement to be 1599 feet; by another measurement it is chained in two parts, and AB found to be 791, and BC 806 feet. Considering the first as of weight 2 and each of the others as of weight 5, what is the adjusted length?

PROBABLE ERRORS.

49. The determination of the weights and probable errors of conditioned observations follows directly from the method of Art. 38, since all such observations can be reduced to independent ones.

Let n denote the number of observations, q the number of unknown quantities involved, and ρ the number of conditional equations. Let g_1, g_2, g_3, etc. be the weights of the several measurements, and v_1, v_2, v_3, etc. the residuals arising from subtracting the observed and adjusted values, and $\Sigma g v^2$ the sum of $g_1 v_1^2 + g_2 v_2^2 + g_3 v_3^2 +$ etc. Then the probable error of an observation whose weight is unity is

$$(111) \qquad r = 0{\cdot}6745 \sqrt{\frac{\Sigma g v^2}{n - q + \rho}},$$

and the probable error of an observation of the weight g_1 is $\dfrac{r}{\sqrt{g_1}}$ (Art. 30). Also if G_s, G_t, etc. be the weights of the adjusted values of s, t, etc. their probable errors are

$$(81) \qquad R_s = \frac{r}{\sqrt{G_s}}, \qquad R_t = \frac{r}{\sqrt{G_t}}, \text{ etc.},$$

to determine which we must first find r, and the weights G_s, G_t, etc. If as in Arts. 44 and 45 all the observations are of equal weight, we have only to make g, g_1, g_2, etc. unity or simply omit them from the formulae. To find the weights G_s, G_t, etc. we have only to form the independent observation equations by elimination of unknown quantities from the conditional equations (Art. 44), and then having multiplied each by the square root of its weight, proceed with the normal equations by the method of Art. 39. A single example will render the whole operation clear to the reader.

We take a case of levelling involving but one conditional equation. There are three points A, B, and C, situated at nearly equal distances apart, but upon different levels. In order to ascertain with accuracy their relative heights, a levelling instrument was set up between A and B, and

CONDITIONED OBSERVATIONS. 95

readings taken upon a rod held at those points, with the results,

On Rod at A, 8·7342 feet—mean of 12 readings,
" " B, 2·3671 " — " 9 "

The instrument was then moved to a point between B and C, and the observations taken

On Rod at B, 5·0247 feet—mean of 7 readings,
" " C, 11·2069 " — " 4 "

Lastly, the level was set up between C and A, and the rods observed

On Rod at C, 0·4672 feet—mean of 5 readings,
" " A, 0·6510 " — " 3 "

It is required to find the adjusted values of these readings and their probable errors, also the most probable differences of level between the points and their probable errors.

First let us arrange these measurements as they would be written in an engineer's level book, and assuming the elevation of A as 0·0, find the heights of the other points.

Sta.	Back Sight.	Fore Sight.	Height of Instrument.	Elevation above A.
$_{12}A$	8·7342		8·7342	0·0
$_7B_9$	5·0247	2·3671	11·3918	6·3671
$_5C_4$	0·4672	11·2069	0·6521	0·1849
A_3		0·6510		0·0011

The number of readings or the *weight* of each sight is placed in the first column preceding and following the name of the station; thus $_7B_9$ denotes that the back sight on B has a weight of 7, and the fore sight one of 9. Regarding the elevation of A as 0, that of B comes out 6·3671 feet, that of C, 0·1849 feet, and on returning to the starting-point we find that A is 0·0011 feet instead of 0 as it ought to be.

CONDITIONED OBSERVATIONS.

If we represent the back sights upon A, B, and C by S, T, and W, and the fore sights upon B, C, and A by X, Y, and Z, the rigorous conditional equation is

$$S + T + W - X - Y - Z = 0.$$

Since this condition is not exactly fulfilled by the observed values, they must be corrected so as to cause the discrepancy of 0·0011 to disappear. To avoid the use of large numbers let us take the measured values as approximate, and represent by s, t, w, etc. the corrections to be applied to S, T, W, etc. Then our conditional equation is

$$s + t + w - x - y - z = -0·0011.$$

The weights of s, t, w, x, y, and z are 12, 7, 5, 9, 4, and 3, corresponding to the weights of the observations to which they are corrections. The most probable values of the unknown quantities which will exactly satisfy the conditional may then be found, either by the method of Art. 47 or by that of Art. 48. The latter is much the shorter and simpler. Leaving the solution as an exercise for the student, we give merely the results, viz.

$$s = -0·00008, \quad t = -0·00014, \quad w = -0·00020,$$
$$x = 0·00011, \quad y = 0·00024, \quad z = 0·00033.$$

Applying these to the observed values we have the adjusted results, viz.

Sta.	Back Sight.	Fore Sight.	Elevation above A.
A	8·73412		0·0
B	5·02456	2·36721	6·36691
C	0·46700	11·20714	0·18433
A		0·65133	0·0

To find the probable errors of these values we must first determine the weights of the adjusted back and fore sights by the method of Art. 39. We give the process in full for those necessary to determine the elevation of B, that is for S and X.

CONDITIONED OBSERVATIONS.

Remembering that s, t, etc. are corrections to the observed values, the observation equations are

$s = 0$ weight 12, $\qquad x = 0$ weight 9,
$t = 0$,, 7, $\qquad y = 0$,, 4,
$w = 0$,, 5, $\qquad z = 0$,, 3,

and the conditional equation is as before

$$s + t + w - x - y - z = -0\cdot0011.$$

Taking from this the value of y in terms of the other unknown quantities, inserting it in the observation equations, and multiplying each by the square root of its weight (Art. 36), we have

$$\sqrt{12}\, s = 0,$$
$$\sqrt{7}\, t = 0,$$
$$\sqrt{5}\, w = 0,$$
$$3\, x = 0,$$
$$\sqrt{3}\, z = 0,$$
$$2s + 2t + 2w - 2x - 2z = -0\cdot0022.$$

From these we have the normal equations (Art. 34),

$$16s + 4t + 4w - 4x - 4z = -0\cdot0044 = A,$$
$$4s + 11t + 4w - 4x - 4z = -0\cdot0044 = B,$$
$$4s + 4t + 9w - 4x - 4z = -0\cdot0044 = C,$$
$$-4s - 4t - 4w + 13x + 4z = 0\cdot0044 = D,$$
$$-4s - 4t - 4w + 4x + 7z = 0\cdot0044 = E,$$

the first being the normal equation for s, the second for t, the third for w, the fourth for x, and the fifth for z.

To determine the weight of s, we place the absolute terms in the normal equations, equal to A, B, C, D, and E, and then solve the equations by any algebraic method. Then the weight of s will be the reciprocal of the coefficient of A in the value of s, and the weight of x will be the reciprocal of the coefficient of D in the value of x (Art. 39), The solution gives

M.

$$s = \frac{3921}{50832} A + \text{terms in } B, C, D, \text{ and } E = -\ 0{\cdot}00008,$$

$$x = \frac{212}{2119} D + \text{terms in } A, B, C, \text{ and } E = \ \ 0{\cdot}00011.$$

Hence the weight of the value of s is

$$G_s = \frac{50832}{3921} = 12{\cdot}96,$$

and the weight of the value of x is

$$G_x = \frac{2119}{212} = 9{\cdot}99.$$

The original weights of the observed values of S and X were 12 and 9. The adjustment has then increased each of these by nearly unity. (In finding the weight of s, if we do not at the same time wish to find the value of s, the terms B, C, D, and E may be made zero, and the numerical work thus shortened.)

We can now find the probable errors. The residuals $v_1, v_2, v_3,$ etc. are in this case our corrections s, t, w, etc.: squaring these and multiplying each square by the weight of its corresponding observation, we obtain

$$\Sigma g v^2 = 0{\cdot}000001079.$$

Since there are six observation equations, six unknown quantities and one conditional equation, we have $n - q + \rho = 1$. The probable error of an observation whose weight is unity, that is of a single reading of the rod, is then

$$r = 0{\cdot}6745 \sqrt{0{\cdot}000001079} = 0{\cdot}000701,$$

and the probable error of any observation is this value divided by the square root of its weight. Hence the probable error of the adjusted reading $S = 8{\cdot}73412$, whose weight we have found to be 12·96, viz.

$$R_s = \frac{r}{\sqrt{G_s}} = \frac{0{\cdot}000701}{\sqrt{12{\cdot}96}} = 0{\cdot}00019,$$

and the probable error of the adjusted reading X whose weight we have found to be 9·99, is

$$R_x = \frac{r}{\sqrt{10}} = 0.00022.$$

The elevation of B is the difference of the readings S and X or

$$B = 8.73412 - 2.36721 = 6.36691.$$

Hence (Art. 41) its probable error is

$$R = \sqrt{R_s^2 + R_x^2} = 0.00029.$$

The adjusted values may then be written

Back sight on $\quad A = 8.73412 \pm 0.00019,$
Fore sight on $\quad B = 2.36721 \pm 0.00022,$
Height of B above $A = 6.36691 \pm 0.00029.$

In the same way the probable errors of the other adjusted measurements and the probable error of the elevation of C may be deduced.

Problems. 1. In the preceding example what is the probable error of the observed readings 8·7342 and 2·3671, and the probable error of the elevation 6·3671?

2. The three angles of a triangle were measured as follows, all having the same weight,

$$x = 98° \, 17' \, 22'', \qquad y = 70° \, 9' \, 56'', \qquad z = 11° \, 32' \, 52''.$$

What are the adjusted angles and their probable errors?

Ans. $x = 98° \, 17' \, 18''{\cdot}677 \pm 1''{\cdot}007.$

CHAPTER V.

THE DISCUSSION OF PHYSICAL OBSERVATIONS.

50. In the preceding pages we have given the methods of adjusting and comparing such simple observations as arise in the common practice of the civil engineer. It is, indeed, in the delicate measurements of higher Geodesy and Astronomy that these methods of combination find their most extended application, but it would interfere with the plan of this book to introduce examples of such cases. One or two applications, which we have not yet exemplified, are however of such great use, not only to engineers, but to all who cultivate the physical sciences, that we shall devote a few pages to their development and illustration.

The Deduction of Empirical Formulæ.

51. Observations are frequently made to discover the laws which govern phenomena. It is one of the most happy applications of the method of least squares, that it is often able to determine from such observations the relations between the observed quantities, and to deduce formulæ expressing these relations in a convenient form for use. When such formulæ merely represent the particular observations from which they are deduced, they are called *empirical*, but when they are applicable to all phenomena of the same class they become *physical laws*. The discussion of observations to deduce empirical relations thus often leads to the discovery of important laws by which our physical theories are extended and improved.

To illustrate, let us suppose that the law of falling bodies is unknown, and that in order to discover the relation between the time of falling and the space passed over, we construct an apparatus by which a body can be allowed to fall certain given distances, and the times of its descent be accurately registered. Suppose that with this apparatus the following observations are made at New York.

1. A body starts from rest and falls 10 feet in 0·788 seconds.
2. „   „   „   20 „ 1·115 „
3. „   „   „   30 „ 1·367 „
4. „   „   „   40 „ 1·577 „
5. „   „   „   50 „ 1·763 „

Now what relation or law exists between the space s and the time t? First we observe that as the space increases so does the time, but the first very much faster than the second. The relation between s and t cannot therefore be expressed by such an equation as $s = At$ where A is a constant, but s must depend upon higher powers of t. In like manner we might try in succession the hypotheses that $s = At^2$, $s = At + At^2$, etc. Of course in this simple case the reader will at once recognize the relation which exists. In most cases, however, like those that will be hereafter given, the relation cannot be determined by inspection or even by the trial of hypotheses. In our ignorance, then, of whether s depends upon t, t^2, or t^3, or upon a combination of these, it is best to write

(112) $$s = At + Bt^2 + Ct^3,$$

in which A, B, and C are constants to be determined, and as yet unknown. Then our first observation gives $s = 10$, $t = 0·788$, the second gives $s = 20$ and $t = 1·115$, etc. Substituting then in the assumed formula, we have

1. $10 = 0·788\,A + 0·788^2\,B + 0·788^3\,C,$
2. $20 = 1·115\,A + 1·115^2\,B + 1·115^3\,C.$
    etc.   etc.

There will be as many observations as there are observations each containing the unknown constants A, B, and C. They are then the observation equations (Art. 32) from which we may derive a normal equation (Art. 34) for each of the unknown quantities, and by their solution find the values of A, B, and C. In the case before us this process will give

$$A = 0, \quad B = 16{\cdot}08, \text{ and } C = 0,$$

and hence the assumed formula becomes

$$s = 16{\cdot}08\, t^2.$$

With reference to the five observations this is an empirical formula, but as it is found to represent all such observations made at New York, it is hence also an expression of the law of falling bodies at that place.

52. A very large class of physical phenomena may be represented by the general equation

(112) $\qquad y = A + Bx + Cx^2 + Dx^3 + \text{etc.}$

in which y and x are two related quantities of different kinds and A, B, C, etc. constants. As such we may indicate many of the relations between space and time, space and velocity, volume and temperature, pressure and density, etc. It includes hence many of the empirical formulæ of hydraulics, of the theory of heat, and of other branches of physics. No general rules, however, can be laid down to show when and where it is applicable.

Temperature of the Earth at depths below the surface*. It has long been known that as we descend below the surface of the earth the temperature increases. Geologists formerly supposed that the increase was nearly uniform, and at the rate of about 1 degree Centigrade to every 30 metres in depth. The observation at the artesian well at Grenelle near Paris, showed however that the increase was not at so rapid a rate. The observations there taken were the following†:

---

* From Von Freeden's *Praxis der Methode der Kleinsten Quadrate*, Braunschweig, 1863.
† If F = Fahrenheit degrees and C = Centigrade degrees, $F = 32 + \tfrac{9}{5} C$.
1 meter = 3·28087 *English* feet = 3·28071 *U. S.* feet.

Mean yearly temperature of the surface $= 10°·60$ Centigrade.

1. Temperature at a depth of 28 meters $= 11·71$,,
2. ,, ,, ,, 66 ,, $= 12·90$,,
3. ,, ,, ,, 173 ,, $= 16·40$,,
4. ,, ,, ,, 248 ,, $= 20·00$,,
5. ,, ,, ,, 298 ,, $= 22·20$,,
6. ,, ,, ,, 400 ,, $= 23·75$,,
7. ,, ,, ,, 505 ,, $= 26·45$,,
8. ,, ,, ,, 548 ,, $= 27·70$,,

Let us represent the mean temperature of the surface by T, and the temperature at any depth by t: also let x be any depth. We then assume that the temperature increases with the depth according to the approximate law,

$$t = T + Ax + Bx^2,$$

and we have to find A and B from the above observation. The first one gives $t = 11°·71$ and $x = 28$, or

$$11·71 = 10·6 + 28\,A + 28^2\,B.$$

The second one gives

$$12·90 = 10·6 + 66\,A + 66^2\,B.$$

Transposing the constant term $10°·6$ to the first members, the observation equations may be written,

$$1°·11 = 28\,A + 28^2\,B,$$
$$2·30 = 66\,A + 66^2\,B,$$
$$5·80 = 173\,A + 173^2\,B,$$
$$9·40 = 248\,A + 248^2\,B,$$
$$11·60 = 298\,A + 298^2\,B,$$
$$13·15 = 400\,A + 400^2\,B,$$
$$15·83 = 505\,A + 505^2\,B,$$
$$17·10 = 548\,A + 548^2\,B.$$

We now proceed to form the normal equations; multiplying the first equation by 28, the second by 66, the third by

173, etc., and adding the products, we have the normal equation for A, viz.

$$29599\cdot 23 = 900706\,A + 404557966\,B.$$

Also multiplying the first equation by 28^2, the second by 66^2, etc., and adding the products, we find

$$13068985\cdot 39 = 404557966\,A + 193410001814\,B$$

as the normal equation for B. Combining these two normal equations we have the values

$$A = 0\cdot 042096, \quad B = -\,0\cdot 000020558,$$

and hence the empirical formula is

$$t = 10°\cdot 6 + 0°\cdot 042096\,x - 0°\cdot 000020558\,x^2.$$

Thus by this formula the temperature t at any depth x may be found, x being in meters and t in Centigrade degrees. The negative sign of the term containing x^2 shows that the increase of temperature is not so rapid as the depth. If in this formula we substitute $x = 28$, $x = 66$, etc., and compute the corresponding values of t, the accuracy of the formula may be seen by comparing the computed and the observed temperatures. The following table is such a comparison.

Depth.	Observed temperature.	Calculated temperature.
0 m.	10·60 C.	10·60 C.
28	11·71	11·76
66	12·90	13·29
173	16·40	17·26
248	20·00	19·78
298	22·20	21·32
400	23·75	24·15
505	26·43	26·61
548	27·70	27·49
1000	——	32·13

The close agreement between experience and calculation show that *as far as the above observations are concerned*, the formula is sufficiently accurate. A more recent boring at Speremberg near Berlin, which reached a depth of about 1200 meters, seems to indicate a somewhat faster rate of increase. At a depth of 1000 metres the observed temperature was about 43° C, while that computed by our formula is only 32°·13, and it can hence be regarded only as empirical, and as representing the Grenelle observations. A discussion of the results from this deep boring at Speremberg may be seen in *Nature*, Oct. 21, 1875, where it is claimed that after a depth of about 5000 feet the temperature is uniform, and probably not more than 49° Centigrade or 120° of Fahrenheit's scale.

Problem. Volume of water at different temperatures. In MÜLLER's *Lehrbuch der Physik*, Vol. II. p. 505, the following are the relative volumes of water at temperatures from 4° C. to 25° as derived from experiments by DESPRETZ:

At 4° Centigrade the volume is 1·
„ 6 „ „ 1·0000309,
„ 8 „ „ 1·0001216,
„ 10 „ „ 1·0002684,
„ 15 „ „ 1·0008751,
„ 20 „ „ 1·0017900,
„ 25 „ „ 1·0029300.

Required a formula to represent these experiments.

Let V represent the volume at any temperature t. Then by the same process as that of the preceding example, we find

$$V = 1 - 0\cdot000061045\, t + 0\cdot0000077183\, t^2 - 0\cdot00000003734\, t^3.$$

Although agreeing tolerably well with the observations this formula does not hold for temperatures much higher than 25° C. From 25° to 50° MÜLLER gives

$$V = 1 - 0\cdot000065415\, t + 0\cdot0000077587\, t^2 - 0\cdot00000003541\, t^3.$$

It appears then that the law connecting the volume and temperature of water is not yet understood, and hence that the above formulæ are merely empirical.

53. Another large class of phenomena may be represented by the general equation

106 THE DISCUSSION OF PHYSICAL OBSERVATIONS.

(114) $\quad y = A + B_1 \sin \dfrac{360°}{m} x + B_2 \cos \dfrac{360°}{m} x$

$\qquad + C_1 \sin \dfrac{360°}{m} 2x + C_2 \cos \dfrac{360°}{m} 2x + $ etc.

in which as x increases y passes though repeating cycles. As such may be mentioned the variation of temperature throughout the year, the changes of the barometer, the ebb and flow of the tides, the distribution of heat on the surface of the earth depending on latitude, and in fact all phenomena which repeat themselves like the oscillations of a pendulum. The letters A, B_1, B_2, etc. represent constants which are to be found from the observations by the method of least squares, while m is the number of equal parts into which the whole cycle is divided, and must be taken in terms of the same unit as x. If the several cycles are similar and regular, only the first three terms are required to represent the variation, if not, additional terms are necessary.

Declination of the magnetic needle. The magnetic needle has always in the New England States since the earliest records pointed to the west of true north. Its declination has moreover been slowly changing. Numerous observations from all parts of the country have been collected and carefully discussed by Mr SCHOTT of the U. S. Coast Survey*, from which it has been established that the declination passes through a recurring cycle of about 250 or 300 years. If we divide a horizontal line into equal parts representing years, and erect verticals giving the declination at various times, we shall have a curve, such as is roughly represented

Fig. 14.

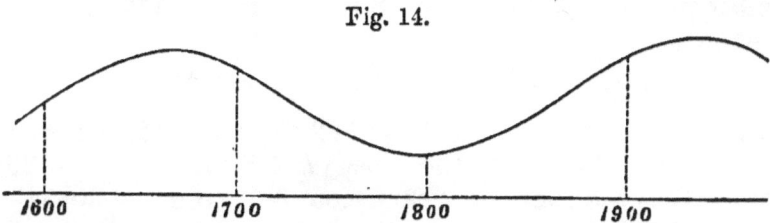

in the annexed figure. Thus in Connecticut the westerly declination was in the year 1600 gradually increasing; it

* See *U. S. Coast Survey Reports*, 1855, 1858 and 1859.

attained a maximum of 11° or 12° about 1675, then again decreased until it reached a minimum of 5° about 1800, and has since been steadily but not uniformly increasing.

If then we have observations enough to determine the equation of this curve, we shall not only have an empirical formula representing those observations, but an expression of the law of the change. At any place where observations extending over fifty years exist, an approximate formula may be found.

At Hartford, Conn., we have the following records:

In 1786 the declination was 5° 25′ W = + 5°·42,
„ 1810 „ „ 4 46 „ = 4 ·77,
„ 1824 „ „ 5 45 „ = 5 ·75,
„ 1828 „ „ 6 3 „ = 6 ·05,
„ 1829 „ „ 6 3 „ = 6 ·05,
„ 1859 „ „ 8 4 „ = 8 ·07.

We assume the expression

$$D = A + B_1 \sin \frac{360°}{m} t + B_2 \cos \frac{360°}{m} t,$$

as a sufficient approximation to the law of the declination, in which D is the value of the declination at any time t, t being the number of years counted from an assumed epoch, say the year 1830, m the number of years in the cycle, and A, B_1 and B_2 constants to be determined. The value of m is not exactly known, and must be adopted from observations at other places, or be determined by trial. Mr SCHOTT finds its most probable value to be 288 years for Hartford. Then our equation is

$$D = A + B_1 \sin 1·25\, t + B_2 \cos 1·25\, t.$$

The first observation is $D = 5°·42$ for $t = -44$, the second $D = 4°·77$ for $t = -20$, the last $D = 8°·07$ for $t = +29·6$, t being the number of years counted from 1830 as an origin, and hence negative for the years preceding that epoch and positive for those following. The first observation equation is then

$$5°·42 = A - 0·819\, B_1 + 0·574\, B_2,$$

and the five others are similar. (The *sign* of the sine and cosine of a negative arc should be carefully regarded.)

From the six observation equations we form and solve the normal equations, and find
$$A = 8°{\cdot}60, \quad B_1 = 2°{\cdot}54, \quad B_2 = -2°{\cdot}54,$$
hence our formula is
$$D = 8°{\cdot}60 + 2°{\cdot}54 \sin 1°{\cdot}25\, t - 2°{\cdot}54 \cos 1°{\cdot}25\, t,$$
which agrees very closely with the observations. This equation may be discussed like that of any algebraic curve, and the times and values of the maximum and minimum declination found. Thus, according to the formula, the last minimum at Hartford was $5°{\cdot}01$ for $t = -34$, or in the year 1796; the next maximum will be $12°{\cdot}19$ for $t = 110$, or in the year 1940. The formula therefore extends the law of the variation with a fair degree of accuracy to times considerably preceding and following the observations. As indicated in Part II. it may be placed under the forms
$$D = 8°{\cdot}60 + 3°{\cdot}59 \sin (1°{\cdot}25\, t - 45°),$$
$$D = 8°{\cdot}60 - 3°{\cdot}59 \cos (1°{\cdot}25\, t + 45°),$$
either of which is more convenient for discussion than that given above; and the latter of which is the form given by SCHOTT*.

54. Other general empirical and theoretical formulæ such as
$$y = A + Bx^{\frac{1}{2}} + Cx^{\frac{3}{2}} + \text{etc.}$$
$$y = A + B \sin mx + C \sin^2 mx + \text{etc.,}$$
are sometimes used in discussing physical phenomena. Rarely also coefficients such as A^2, B^2, etc. are introduced, but their determination is generally laborious (see Part II. Art. 59). Exactly what formula will apply to a given set of observations, so as to agree well with them, and at the same time be of use in other similar cases, can only be determined from theoretical considerations or by trial. The computer must, first, from his knowledge of physical laws, assume such an

---

* *U. S. Coast Survey Report*, 1859, p. 298, or *Am. Journal of Science*, 1860, p. 335, where declination formulæ for many other places are given.

expression as seems most plausible, then deduce the constants by the method of least squares, and test them by comparing the observed with the calculated results. If the agreement is not sufficiently close, he must assume another expression and try again. In this way hypotheses may be tested, and often important physical laws be discovered.

The precision of such empirical formulæ may be ascertained not only by their agreement with the observations, but by computing the probable errors of the constants, which the methods of least squares has deduced (Art. 38). The probable errors of the calculated results may also be found, but for such computations we must refer the reader to the larger and more complete treatises upon the subject, a list of which will be found in the Appendix.

Problem. The following observations have been made in different parts of the earth, upon the length of the second's pendulum.

Place.	Latitude.	Length of Pendulum in meters.
Peru	0°	0·990564
Petit Goave	18° 27′	0·991150
Paris	48 24	0·993867
St Petersburg	58 15	0·994589
Lapland	67 4	0·995325

Find a formula representing these observations, observing that according to the theory of the variation of gravity it must be of the form

$$l = A + B \sin^2 \phi,$$

in which ϕ is any latitude, and l the corresponding length of the second's pendulum.

Ans. $l = 0^m \cdot 990555 + 0^m \cdot 005679 \sin^2 \phi.$

BOWDITCH, in his Commentary on LAPLACE's *Mécanique Céleste,* finds a formula for the length of the pendulum by using *fifty-two* observations. The student who has solved the above problem will have some idea of the patience and labour required in such processes, and will be interested in looking over BOWDITCH's admirable discussion, which is in Vol. II. p. 481 of the above-cited work. The formula which he adopts as the best is

$$l = 39 \cdot 01307 + 0 \cdot 20644 \sin^2 \phi,$$

in which the constants and l are in inches, while those of our problem are in meters*. IVORY's interesting papers in the *London Philosophical Magazine* for 1826 may also be referred to in this connection.

Probability of Errors.

55. In Arts. 12 and 13 we have stated a property of the probability curve, and given a table of the probabilities of errors, which has many interesting applications in various branches of science. That table gives the areas of the curve on both sides of the axis of Y, corresponding to successive numerical values of hx, h being the constant measure of precision and x any error, or, more strictly, it gives the number of errors comprised between the limits $-x$ and $+x$, when the whole number of errors is unity. As the frequency of an error is proportional to its probability (Art. 5), the numbers in the columns P' denote also the probabilities that an error will be comprised between the assumed limits. Thus if in the figure representing the probability curve, we lay off the positive error OM equal to the negative error OM, and draw the ordinates MC and MC, the area $MCACM$ is (if the total area be unity) a fraction expressing the probability that an error will lie between the limits $-OM$ and $+OM$, or be numerically less than OM. So also if the area $PBAPB$ be 0·5, OP is the probable error (Art. 16), or the error such that the probability of an error less or greater than it is ½. If we were to lay off an abscissa positively and negatively equal to three times OP, the corresponding area would be 0·957, and the probability that an error taken at

---

* 1 meter = 39·37044 *English* inches = 39·36852 *U. S.* inches. I do not know whether the above formula is in English or United States inches. This annoying difference between our own measures and those of England is one of the strongest reasons why we should throw both of them overboard and adopt a reasonable system.

Fig. 1.

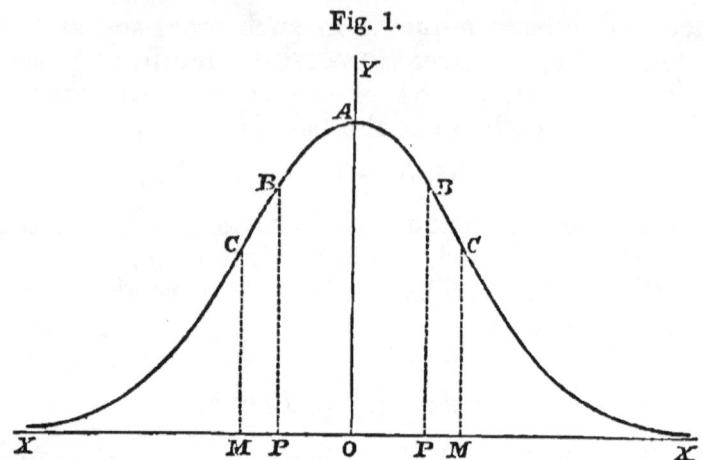

random would be less than $3.\overline{OP}$ is $\frac{957}{1000}$, and that it would be greater $\frac{43}{1000}$; so that it would be a wager of 957 to 43, or of nearly 22 to 1, that any error is less than three times the probable error.

In discussing the probabilities of errors and the accuracy of sets of observations, it is hence convenient to have a table giving the numbers P' in terms of r instead of l. The following is such a table, which we have taken from CHAUVENET's *Treatise on Least Squares*, and which gives these areas or probabilities, P' corresponding to successive numerical values of $\frac{x}{r}$, x being any error and r the probable error. To illustrate its use let us consider some practical examples.

1. Suppose that we have measured an angle with two instruments A and B, and find for the results

with A $37°\ 42'\ 13''\!\cdot\!9 \pm 0''\!\cdot\!4$,

with B $37°\ 42'\ 13''\!\cdot\!8 \pm 0''\!\cdot\!2$,

the probable error of the first being twice that of the second, and hence its precision one-half as great (see Fig. 2, Art. 16). The probability that the result A does not differ from the

112 THE DISCUSSION OF PHYSICAL OBSERVATIONS.

Probability of Errors. (115). $P' = \dfrac{2}{\sqrt{\pi}} \displaystyle\int_0^t e^{-t^2} dt, \quad t = hx = 0{\cdot}4769 \dfrac{x}{r}.$

$\dfrac{x}{r}$	P'	$\dfrac{x}{r}$	P'	$\dfrac{x}{r}$	P'	$\dfrac{x}{r}$	P'
0·00	0·00000	1·20	0·58171	2·40	0·89450	3·60	0·98482
0·04	·02152	1·24	·59705	2·44	·90019	3·64	·98588
0·08	·03228	1·28	·61205	2·48	·90562	3·68	·98691
0·12	·06451	1·32	·62679	2·52	·91082	3·72	·98786
0·16	·08594	1·36	·64102	2·56	·91578	3·76	·98872
0·20	0·10731	1·40	0·65498	2·60	0·92051	3·80	0·98962
0·24	·12860	1·44	·66858	2·64	·92503		
0·28	·14980	1·48	·68184	2·68	·92934		
0·32	·17088	1·52	·69474	2·72	·93344		
0·36	·19185	1·56	·70729	2·76	·93734		
0·40	0·21268	1·60	0·71949	2·80	0·94105	4·00	0·99302
0·44	·23336	1·64	·73134	2·84	·94458		
0·48	·25388	1·68	·74285	2·88	·94793		
0·52	·27421	1·72	·75400	2·92	·95111		
0·56	·29436	1·76	·76481	2·96	·95412		
0·60	0·31430	1·80	0·77528	3·00	0·95698	5·00	0·99926
0·64	·33402	1·84	·78542	3·04	·95968		
0·68	·35352	1·88	·79522	3·08	·96224		
0·72	·37277	1·92	·80469	3·12	·96466		
0·76	·39178	1·96	·81383	3·16	·96694		
0·80	0·41052	2·00	0·82266	3·20	0·96910		
0·84	·42899	2·04	·83117	3·24	·97114		
0·88	·44719	2·08	·83936	3·28	·97306		
0·92	·46509	2·12	·84726	3·32	·97486		
0·96	·48270	2·16	·85480	3·36	·97573		
1·00	0·50000	2·20	0·86216	3·40	0·97817		
1·04	·51699	2·24	·86917	3·44	·97967		
1·08	·53366	2·28	·87591	3·48	·98109		
1·12	·55001	2·32	·88237	3·52	·98237		
1·16	·56602	2·36	·88857	3·56	·98360	∞	1·00000

true value of the angle by $0''\!\cdot\!4$ is $\frac{1}{2}$, and the probability that the result B does not differ by $0''\!\cdot\!2$ is also $\frac{1}{2}$. Now what are the respective probabilities that these results are correct within $0''\!\cdot\!1$? We take then $x = 0''\!\cdot\!1$, and have

for A, $\dfrac{x}{r} = \dfrac{0\cdot 1}{0\cdot 4} = 0\cdot 25$, and for B, $\dfrac{x}{r} = \dfrac{0\cdot 1}{0\cdot 2} = 0\cdot 5$;

hence from the table

for $\dfrac{x}{r} = 0\cdot 25$, $P' = 0\cdot 134$, and for $\dfrac{x}{r} = 0\cdot 5$, $P' = 0\cdot 264$.

The probability that the result A is within $0''\!\cdot\!1$ of the truth is $\dfrac{134}{1000}$, and that B is within the same limit $\dfrac{264}{1000}$ or nearly twice as much. Hence for A we could wager 134 to 866 or 1 to 5 that the result was within $0''\!\cdot\!1$ of the truth, and for B 264 to 736 or 1 to 2·8 that such was the case. Again, what are the respective probabilities that these results are within $0''\!\cdot\!6$ of the truth? We have

for A, $\dfrac{x}{r} = \dfrac{0\cdot 6}{0\cdot 4} = 1\cdot 5$, and for B, $\dfrac{x}{r} = \dfrac{0\cdot 6}{0\cdot 2} = 3\cdot 0$,

and then from the table

for $\dfrac{x}{r} = 1\cdot 5$, $P' = 0\cdot 688$, and for $\dfrac{x}{r} = 3$, $P' = 0\cdot 957$.

The probabilities are then $\dfrac{688}{1000}$ and $\dfrac{957}{1000}$ respectively; and we could afford to lay a wager of 688 to 312 or of about 2 to 1 that A is within $0''\!\cdot\!6$ of the truth, and one of 957 to 43 or of more than 22 to 1 that B is within the same limit.

2. A series of 64 observations upon an angle gives for the probable error of a single observation $1''\!\cdot\!5$. What is the probability that the mean is correct within $0''\!\cdot\!75$?

The probable error of the mean will be $\dfrac{1\cdot 5}{\sqrt{64}} = 0''\!\cdot\!185$.

Then $\dfrac{x}{r} = \dfrac{0.75}{0.185} = 4.05$, for which $P' = 0.994$.

The required probability is hence $\dfrac{994}{1000}$, and it is a wager of 994 to 6 or of 166 to 1 that the mean is within $0''\!\cdot\!75$ of the true value.

3. An angle is measured with an instrument graduated to $1'$. The error which is liable to occur in a single measurement (that is, the probable error r) is $45''$. How many observations are necessary in order that it shall be a wager of 9 to 1 that the mean is within $5''$ of the truth?

A wager of 9 to 1 corresponds to a probability of $\dfrac{9}{10}$. For $P' = 0.9$ we find from the table $\dfrac{x}{r} = 2.44$, in which r, being the probable error of the average, is equal to $\dfrac{45''}{\sqrt{n}}$. Then we have

$$2.44 = \dfrac{x}{r} = \dfrac{x\sqrt{n}}{45} = \dfrac{5\sqrt{n}}{45}, \text{ or } \sqrt{n} = 21.96, \text{ and } n = 482,$$

and hence 482 observations are necessary.

Problems. 4. The average of several observations gives for the value of an angle $33°\ 17'\ 30''\!\cdot\!8 \pm 0''\!\cdot\!3$. What wager can we afford to lay that $33°\ 17'\ 30''\!\cdot\!8$ is within $1''$ of the true value? *Ans.* 39 to 1.

5. A line is measured 500 times. If the probable error of each observation is 0·6 centimeters, how many errors will be less than 1 centimeter and greater than 0·4 centimeters?

6. An angle is measured by an instrument graduated to quarter degrees, the probable error of a single reading being 12 minutes: how many observations are necessary that it may be a wager of 5 to 1 that the mean is within one minute of the truth? *Ans.* 22 observations.

7. The length of a line was estimated by 16 persons of equal skill. If the probable error of each is 4 centimeters,

THE DISCUSSION OF PHYSICAL OBSERVATIONS. 115

what is the probability that the mean of their guesses is within 2 centimeters of the truth?

56. The preceding examples illustrate the great value of large numbers of observations even when made with poor instruments, provided only that no constant cause of error exists (Art. 2). The preceding table is thus of the greatest assistance in discussing statistics of social science, even when those statistics are confessedly inaccurate, provided only that they are numerous and taken by unprejudiced observers. It is however not the place here to give examples of such applications, and we would refer the reader desirous of looking up such investigations to the admirable work of QUETELET (No. 17 in our list of literature in Art. 64 of the Appendix), in which numerous interesting examples of such discussions may be seen.

The principles of probability of error here set forth are, it must be borne in mind, only applicable after all constant errors (Art. 2), and all mistakes (Art. 3) have been eliminated from the numerical results. If a single cause of constant error exists it may sometimes be detected by a comparison of the results with those obtained by a more accurate instrument. The following illustration may in this connection be interesting.

Suppose that an angle is laid out with very accurate instruments and tested in many ways so that its true value may be regarded as exactly $90°$. Let 25 observations be taken upon it with a transit whose accuracy we wish to test, and let the mean of those measurements be $89° \, 59' \, 57'' + 0''\!\cdot\!8$. Then we see it is extremely probable that a constant error of about $-3''$ exists in the instrument. To find the numerical expression of this probability let us suppose that the true value of the angle was unknown, and let us ask the probability that the mean is within $2''$ of the truth. Then for $\dfrac{x}{r} = \dfrac{2}{0\!\cdot\!8} = 2\!\cdot\!5$ we find $P' = 0\!\cdot\!908$, so that it is a wager of 908 to 92 or of almost 10 to 1 that the mean is between the limits $89° \, 59' \, 55''$ and $89° \, 59' \, 59''$. Hence since the angle is known to be $90°$, it must be the same probability and the same wager that there is a constant error lying between the

8—2

limits $-1''$ and $-5''$. So also if we take $x = 3''$ we can show that it is a wager of 39 to 1 that there is a constant error between $0''$ and $-6''$.

Problem. The capacity of a certain large vessel is unknown: 1600 persons guess at the number of gallons of water which it will hold and the average of their guesses is 289 gallons. The vessel was then measured by a committee and found to hold 297 gallons. If we regard the probable error of a single guess as 50 gallons and also consider it impossible that there can be any constant source of error in guessing, what is the probability that the committee made an error in their measurement of between 3 and 13 gallons?

$$Ans. \quad \frac{993}{1000}, \text{ or a wager of 142 to 1.}$$

The Rejection of Doubtful Observations.

57. It not unfrequently happens that in a set of measurements there are certain values which seem to be so much at variance with the majority, that the observer rejects them in adjusting the results. This procedure however, unless governed by proper rules, is a dangerous one and not to be recommended. A conscientious observer having conscientiously made several series of measurements will give each its proper weight (Art. 28) and deduce the most probable result, and give to it the confidence which its probable error shows that it deserves. The too common practice of taking twenty measurements (for instance, readings of a levelling rod) and then throwing away all except two or three which happen to agree, is one which cannot be strongly enough condemned. After making such measurements and eliminating all known *constant errors* (Art. 2), no results except those which are unquestionable *mistakes* (Art. 3) should be rejected. All remaining discrepancies will then fall under the class of *irregular errors* (Art. 3), and the adjustment of such observations should be made in accordance with the principles governing them, the methods for which we have presented in the preceding Chapters.

We must mention, however, that for very delicate and precise observations such as sometimes arise in Astronomy, the principles of probability itself furnish a means of determining whether or not a given observation may be rejected. The discovery of this important criterion is due to Prof. PEIRCE of Harvard University, and presented in the *Cambridge Astronomical Journal* for 1856. The student wishing to inform himself concerning its theory and application, may consult the original paper in that Journal or the works Nos. 28 and 34 quoted in the list of literature given in Art. 64 of our Appendix. The use of this criterion is not necessary except in some very accurate astronomical investigations.

Concluding Remarks.

58. The student who has carefully read the foregoing pages and has verified the examples and solved the problems presented, will have acquired a fair knowledge of the principle of Least Squares and of its simpler applications to engineering practice, and will be prepared to study the more complete theory of the subject with interest and profit. The second part of this book will afford him an introduction to the theory, which has been prepared with especial reference to the needs of beginning students. He will also be prepared to take up the adjustment of Geodetic and Astronomical observations involving large numbers of observations, and as excellent books for study and consultation in this connection, we would refer him particularly to the works of CHAUVENET, VON FREEDEN, DIENGER and HELMERT, whose titles will be found in the list of literature appended in Art. 64. The brief history of the origin and development of the science given in the Appendix will also prove suggestive in directing his further studies.

We take occasion to again call attention to the fact that the formulæ given for probable errors are all based upon the supposition that the number of measurements is large. Hence in using different formulæ a perfect agreement in the results is not to be expected, unless sufficient observations have been taken to exhibit the several errors in proportion to their respective probabilities, and this would require a very large number. We would also again mention that the whole theory is based upon the supposition that only accidental or irregular errors (Art. 4) affect the measurements.

Although we have only given methods for the adjustment of observations involving equations of the first degree they are sufficient for the discussion of all cases; for, by the process of Art. 59 (Appendix), equations of higher degrees can always be reduced to linear equations, and the observations thus be brought under the rules which we have presented.

PART II.

THE THEORY OF LEAST SQUARES AND PROBABLE ERRORS.

CHAPTER VI.

DEDUCTION OF THE FUNDAMENTAL PRINCIPLES.

1. In Part I. we have presented the rules and methods for the adjustment and comparison of ordinary observations and have illustrated their application by numerous practical examples and problems. We shall now take up the subject from a more mathematical point of view, and give demonstrations of all the formulæ which have been there employed without proof, and also discuss the method in a more complete and algebraic manner. While the previous chapters have been written with more especial reference to the wants of practical computers, the following pages will be designed to meet the requirements of students who wish to acquire a tolerably thorough knowledge of the theory of the subject. Although each Part may be read independently, each is in fact a supplement to the other, and hence in order to render reference easy, we use the same numbers to mark the corresponding paragraphs and formulæ.

Since all measurements even when made with the utmost precision give discordant results, all of which cannot be true, it is evident that we can never be sure that we have found the absolutely true value of a quantity, which has been the object of measurement. In combining then such measures or observations, we seek a method which shall furnish us with the most advantageous or *most probable* result, that is, a result which (as far as our observations go) we can regard as the nearest approximation to the true value. Such is the method of Least Squares, the theory of which we are to develope. Further, as we cannot regard our adjusted result as absolutely true, we must also establish what degree of confidence it is entitled to, and this involves the theory of probable errors. The first treats then of the *adjustment*, the second of the *comparison* of observations.

When we measure several times a quantity and obtain discordant results, we recognise that each measurement is probably incorrect. The difference between the true and an observed value is called an *error;* it is taken as positive if the true value exceeds, and negative if it is less than the observed one. Since the true value cannot be exactly determined, these errors can never be definitely known, nevertheless they can be made the subject of mathematical investigation.

2. *Constant errors* which always affect our observations by the same amount and whose causes are understood, are no longer errors, as they may be always eliminated from our numerical results. They are not the errors which we are to discuss.

3. *Mistakes*, whose distinguishing feature is that they "admit of conjectural correction," are also not included among the errors of which we are to treat.

4. *Accidental or irregular errors* are those discrepancies which remain after all constant errors and mistakes have been eliminated, and are hence those produced by irregular and varying causes whose degree or manner of action cannot be estimated. The word *error*, as used in this book, means therefore "discordance, of which the cause is unknown." These errors being produced by many causes, all unknown as to their laws, are governed by the principles of probability.

Probability.

5. We give therefore, by way of introduction, the definition and some of the first principles of probability.

DEDUCTION OF THE FUNDAMENTAL PRINCIPLES. 121

The word *probability* as used in mathematical reasoning means a number less than unity, which is the ratio of the number of ways in which an event may happen or fail to the total number of possible ways. Thus if an event may happen in a ways and fail in b ways, and each of these ways is equally likely to occur, the probability of its happening is $\frac{a}{a+b}$, and the probability of its failing is $\frac{b}{a+b}$. Thus probability is always expressed by an abstract fraction, and is a numerical measure of the degree of confidence which we have in the happening or failing of an event. If the fraction is 0 it denotes impossibility, if $\frac{1}{2}$ it denotes that the chances are equal for and against its happening, and if it is 1 the event is certain to happen.

6. Hence unity is the mathematical symbol for *certainty*. And since an event must either happen or not happen, the sum of the probabilities of happening and failing is unity. Thus if P be the probability that an event will happen, $1 - P$ is the probability of its failing.

7. If an event may happen in a ways and also in a' ways and fail in b ways, the probability of its happening is, by Art. 5, $\frac{a+a'}{a+a'+b}$; and since this is the sum of the probability of happening in a ways and of that of happening in a' ways, it follows that if an event may happen in different independent ways the probability of its happening is the sum of the separate probabilities.

8. Let us now ask the probability of the concurrence of two independent events. Let the first be able to happen in a_1 ways and fail in b_1 ways, and the second happen in a_2 and fail in b_2 ways. Then there are for the first event $a_1 + b_1$ possible cases, and for the second $a_2 + b_2$: and each case out of the $a_1 + b_1$ cases may be associated with each case out of the $a_2 + b_2$ cases, and hence there are for the two events $(a_1 + b_1)(a_2 + b_2)$ compound cases each of which is equally likely to occur. In $a_1 a_2$ of these cases both events happen, in $b_1 b_2$ both fail, in $a_1 b_2$ the first happens and the second fails, and in $a_2 b_1$ the first fails and the second happens. Thus we have for two independent events

Probability that both happen ....
$$= \frac{a_1 a_2}{(a_1 + b_1)(a_2 + b_2)},$$

Probability that both fail ....
$$= \frac{b_1 b_2}{(a_1 + b_1)(a_2 + b_2)},$$

Probability that the first happens and the second fails
$$= \frac{a_1 b_2}{(a_1 + b_1)(a_2 + b_2)},$$

Probability that the first fails and the second happens
$$= \frac{a_2 b_1}{(a_1 + b_1)(a_2 + b_2)},$$

and the sum of these is unity since one of the four events is *certain* to occur. Now considering each event alone the probability of the first happening is $\frac{a_1}{a_1 + b_1}$, and of the second $\frac{a_2}{a_2 + b_2}$, and since

$$\frac{a_1 a_2}{(a_1 + b_1)(a_2 + b_2)} = \frac{a_1}{a_1 + b_1} \times \frac{a_2}{a_2 + b_2},$$

we have established the important principle, that the probability of the concurrence of several independent events, is equal to the product of the separate probabilities.

Thus if there be four events and P_1, P_2, P_3 and P_4 be the respective probabilities of happening, the probability that all the events will happen is $P_1 P_2 P_3 P_4$, and the probability that all will fail is $(1 - P_1)(1 - P_2)(1 - P_3)(1 - P_4)$. The probability that the first happens and the other three fail is $P_1(1 - P_2)(1 - P_3)(1 - P_4)$; and so on.

9. If there be several events whose separate probabilities are known, what is the case most likely to occur? Let P be the probability of the happening of an event in one trial and Q the probability of its failing so that $P + Q = 1$: and let there be n such events. Then by the preceding Art. the probability that all will happen is

DEDUCTION OF THE FUNDAMENTAL PRINCIPLES. 123

P^n; the probability that one *assigned* event will fail and $n-1$ happen is $P^{n-1}Q$, and since this may occur in n ways the probability that one will fail and $n-1$ happen is $nP^{n-1}Q$. Similarly the probability of two assigned events failing and $n-2$ happening is $P^{n-2}Q$, and since this may be done in $\dfrac{n(n-1)}{2}$ ways*, the probability that 2 out of the whole number will fail and $n-2$ happen is $\dfrac{n(n-1)}{2}P^{n-2}Q^2$. If then $(P+Q)^n$ be expanded by the binomial formula, thus,

$$(P+Q)^n = P^n + nP^{n-1}Q + \frac{n(n-1)}{1.2}P^{n-2}Q^2 + \ldots$$
$$+ \frac{n(n-1)(n-2)\ldots(n-m+1)}{1.2.3\ldots m}P^{n-m}Q^m + \text{etc.},$$

the first term is the probability that all will happen, the second that $n-1$ will happen and 1 fail, and the $m+1^{\text{th}}$ term is the probability that $n-m$ will happen and m fail.

To determine then the most probable case we have only to find the term in this series which is the greatest. If we consider that n coins be thrown, $P = Q = \tfrac{1}{2}$, and the series becomes

$$(\tfrac{1}{2})^n + n(\tfrac{1}{2})^n + \frac{n(n-1)}{1.2}(\tfrac{1}{2})^n + \frac{n(n-1)(n-2)}{1.2.3}(\tfrac{1}{2}n) + \ldots$$
$$+ \frac{n(n-1)(n-2)\ldots(n-m+1)}{1.2.3\ldots m}(\tfrac{1}{2})^n + \ldots$$

in which if n is even the middle term is the greatest, and if n is odd there are two equal middle terms greater than any other. Thus if $n = 9$, the series is

$$\frac{1}{512} + \frac{9}{512} + \frac{36}{512} + \frac{84}{512} + \frac{126}{512}$$
$$+ \frac{126}{512} + \frac{84}{512} + \frac{36}{512} + \frac{9}{512} + \frac{1}{512},$$

* See TODHUNTER'S *Algebra for Schools and Colleges*. London, 1870, p. 288, p. 455.

and hence if 9 coins be thrown $\frac{1}{512}$ is the probability that all will be heads, $\frac{9}{512}$ the probability that 8 will be heads and 1 tail, and so on. Since one of these 10 groups must happen the sum of the series is unity. The most probable group is that whose probability is the greatest, and this is the one corresponding to $\frac{126}{512}$, viz. 5 heads and 4 tails, or 4 heads and 5 tails.

If then we have several events either simple or compound we recognise that *the most probable case is that whose mathematical probability is the greatest.* Hence if in given sets of observations various adjusted values of the measured quantity have different probabilities, the one to be chosen is that which has the maximum probability (Arts. 11, 15).

Law of the Probability of Error.

10. Although it would seem at first sight that accidental errors of observation could hardly be made the subject of mathematical reasoning, yet the very fact of their irregularity brings them under the laws of probability. Moreover we recognise that they must be subject to the following fundamental laws of arrangement: 1st, Small errors are more frequent than large ones; 2nd, Positive and negative errors (that is, measurements greater and less than the true value) are equally probable, and hence in a large number of observations are equally frequent; 3rd, Very large errors do not occur, so that in every set of observations there is a limit l, such that all the positive errors are included between 0 and $+l$ and all the negative ones between 0 and $-l$.

These are the three fundamental axioms which, in connection with the principles of probability, are the foundation of all our following reasoning.

11. Therefore, different errors are not equally probable; for in a large number of observations a small error occurs more frequently than a large one, and hence has a greater

DEDUCTION OF THE FUNDAMENTAL PRINCIPLES. 125

probability, and an error greater than the limit l has a probability of 0 or is impossible. Hence the probability of an error is a function of that error, so that calling x any error and y its probability, the law of probability of error is represented by the equation

(1) $$y = f(x),$$

and will be determined if we can find the form of $f(x)$.

Although practically there is a limit in the graduation and use of instruments by which x can have only definite numerical values (thus if an observer reads a theodolite to 10″, the values of x can only differ by 10″ or some multiple of 10″), we must in our mathematical treatment regard x as a continuous variable. This is evidently perfectly allowable, since as the precision of our observations increases the successive values of x are separated by smaller and smaller intervals. Taking y then as a continuous function of x the equation (1) represents a curve whose form we are to determine. The axioms of Art. 10 show that its general form

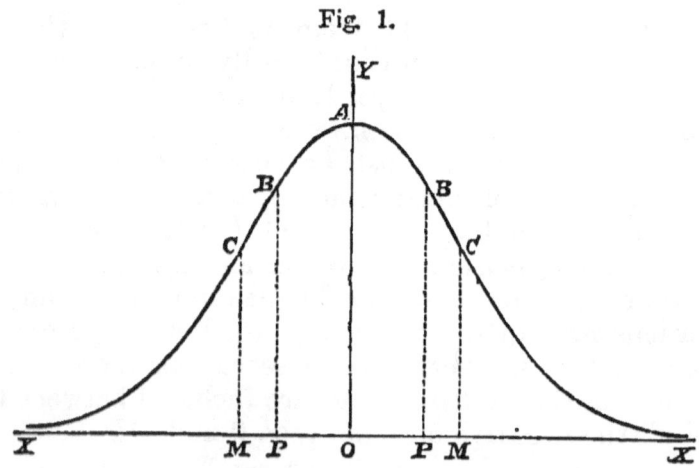

Fig. 1.

must be that of Fig. 1, for by the first small values of x must have the largest probabilities y, by the second $f(x)$ must equal $f(-x)$ or the curve be symmetrical to the axis of Y, and by the third y must be zero for all values of x greater than $\pm l$. This last requirement is one which extremely embarrasses the mathematical treatment of the sub-

126 DEDUCTION OF THE FUNDAMENTAL PRINCIPLES.

ject, since it is impossible to determine a continuous function of x which shall become zero for $x = \pm l$ and also be zero for all values of x from $\pm l$ to $\pm \infty$. But since this limit l can never be accurately assigned we shall extend our limits to $\pm \infty$, and shall determine our curve in such a way that the value of y, although not zero for large values of x, will be so very small as to be practically inappreciable.

For determining the equation of this curve we give two demonstrations, the first for the case of direct observations (Art. 18), and the second for the more general case of indirect ones (Art. 19), both of which lead to the same result. These demonstrations are due to GAUSS (see Art. 65).

11a. Let one and the same quantity, for example the angle M, be measured n times with equal care, giving the values $M_1, M_2, M_3 \ldots M_n$. If z is the *true* value of the angle the errors $(z - M_1), (z - M_2), \ldots (z - M_n)$ have been committed, and their respective probabilities are

(1) $\quad y_1 = f(z - M_1), \; y_2 = f(z - M_2) \ldots y_n = f(z - M_n)$,

that is, if n is the whole number of errors, and the error $(z - M_2)$ occurs n_2 times, $y_2 = \dfrac{n_2}{n} = f(z - M_2)$; and we suppose enough observations to be taken to exhibit the several errors in proportion to their respective probabilities. Now the probability of committing all these errors is by Art. 8 the product of the separate probabilities, or if P denote this quantity

(2) $\quad P = y_1 y_2 y_3 \ldots y_n = f(z - M_1) f(z - M_2) \ldots f(z - M_n)$.

This probability P depends upon z the *true* value of the quantity, *which is unknown*. If we give to z various values we shall have corresponding values of P; and in the impossibility of finding the true value of z we can only find its *most probable* value as given by the n observations, and the most probable value of z is that for which P is a maximum (Art. 9). To find the value of z which makes P a maximum we take the natural logarithm of each member of (2), giving

(3) $\quad \log P = \log f(z - M_1) + \log f(z - M_2) + \ldots$
$$+ \log f(z - M_n),$$

DEDUCTION OF THE FUNDAMENTAL PRINCIPLES. 127

and by the usual rules find the value of z for which $\log P$ is a maximum. Differentiating (3) and placing it equal to zero, we have

(4) $\quad \dfrac{dP}{P} = \dfrac{df(z-M_1)}{f(z-M_1)} + \dfrac{df(z-M_2)}{f(z-M_2)} + \ldots + \dfrac{df(z-M_n)}{f(z-M_n)}$,

or since

(5) $\quad \begin{aligned} df(z-M_1) &= \phi(z-M_1)\, f(z-M_1)\, d(z-M_1), \\ df(z-M_2) &= \phi(z-M_2)\, f(z-M_2)\, d(z-M_2), \end{aligned}$

(in which $\phi(z-M_1)$, $\phi(z-M_2)$, etc. denote new functions of the errors) we have

(4) $\quad \dfrac{dP}{P} = \phi(z-M_1)\, dz + \phi(z-M_2)\, dz + \ldots$
$\qquad\qquad\qquad\qquad + \phi(z-M_n)\, dz = 0.$

Hence the value of z which makes (2) a maximum is that which satisfies the equation

(6) $\quad \phi(z-M_1) + \phi(z-M_2) + \phi(z-M_3) + \ldots$
$\qquad\qquad\qquad\qquad + \phi(z-M_n) = 0.$

And this equation will furnish us with the *most* probable value of z, provided that we can determine the form of the function ϕ.

Now it is universally accepted as an axiom[*] that in direct observations made upon one quantity, the average or arithmetical mean furnishes the most probable result, that is, the most probable value of z in the case under consideration is found by taking the sum of $M_1, M_2 \ldots M_n$ and dividing it by n the number of measurements, or

$$z = \dfrac{M_1 + M_2 + M_3 + \ldots + M_n}{n}.$$

This equation may be written

$$nz = M_1 + M_2 + M_3 + \ldots + M_n,$$

which by transposition becomes

(7) $\quad (z-M_1) + (z-M_2) + (z-M_3) + \ldots + (z-M_n) = 0,$

[*] See Art. 66, p. 195.

that is to say, the arithmetical mean requires that the algebraic sum of the differences or errors shall be zero.

Comparing now (6) and (7) we see that the symbol ϕ means merely the multiplication by a constant, since the value of z must be the same in both; hence

(8) $\phi(z - M_1) + \phi(z - M_2) + \text{etc.}$
$$= k(z - M_1) + k(z - M_2) + \text{etc.},$$

where k is any constant. Inserting in this the values of $\phi(z - M_1)$ etc. from (5), we have

(8*) $\dfrac{df(z - M_1)}{f(z - M_1)\, d(z - M_1)} + \dfrac{df(z - M_2)}{f(z - M_2)\, d(z - M_2)} + \text{etc.}$
$$= k(z - M_1) + k(z - M_2) + \text{etc.}$$

And since this is true for any number of observations, it must be true for one, or two, or three. Hence the corresponding terms in the two members must be equal. If then x be *any* error, and y its corresponding probability, so that

(1) $\qquad\qquad y = f(x),$

we have from (8*)

(9) $\qquad \dfrac{df(x)}{f(x)\,dx} = kx, \quad \text{or} \quad \dfrac{dy}{y} = kx\,dx.$

Integrating this we obtain

(10) $\qquad\qquad \log y = \dfrac{kx^2}{2} + k',$

in which k' is the constant of integration, and the logarithm is taken in the Napierian system. Passing from logarithms to numbers we have

(11) $\qquad\qquad y = e^{\frac{1}{2}kx^2} e^{k'} = ce^{\frac{1}{2}kx^2},$

in which e is the Napierian base, and the constant $e^{k'}$ is placed equal to c. Now y is to be positive, and is to decrease as x increases either positively or negatively, hence the constant k is negative. Placing then $\frac{1}{2}k = -h^2$, we have

(12) $\qquad\qquad y = ce^{-h^2 x^2}$

as the equation of the curve represented in Fig. 1.

DEDUCTION OF THE FUNDAMENTAL PRINCIPLES. 129

This equation satisfies the conditions imposed at the beginning of our investigation, for y is a maximum when $x = 0$, it is symmetrical with respect to the axis of Y, since equal positive and negative values of x give equal values for y; and when x becomes very large y is very small. The constants c and h will be particularly considered hereafter.

11b. Let us now consider the more general case of indirect observations in which the quantities $s, t \ldots z$ are to be determined by measurements of a related quantity $M = f(s, t \ldots z)$. Let n observations be made giving the values $M_1, M_2 \ldots M_n$, and it be required to find from these results the most probable values of $s, t \ldots z$. The measurements being imperfect the results M_1, M_2, etc. cannot be perfectly accurate. Comparing each with the corresponding true value the differences will be errors (Art. 4) which we represent by $x_1, x_2, x_3 \ldots x_n$, each of which is a function of $(s, t \ldots z)$. Then from our equation (1) we have for the several errors

$$(1) \qquad y_1 = f(x_1), \quad y_2 = f(x_2) \ldots y_n = f(x_n).$$

And by Art. 8 the probability of committing the given system of errors is

$$(2) \qquad P = y_1 y_2 y_3 \ldots y_n = f(x_1) f(x_2) \ldots f(x_n).$$

Applying logarithms to this expression it becomes

$$(3) \qquad \log P = \log f(x_1) + \log f(x_2) + \ldots + \log f(x_n).$$

Now the most probable values of the unknown quantities $s, t \ldots z$ are those which render P a maximum (Art. 9), and hence the derivative of P with respect to each of these variables must be equal to zero. Indicating the differentiation we have then the following equations:

$$(4) \quad \begin{aligned} \frac{dP}{P\,ds} &= \frac{df(x_1)}{f(x_1)\,ds} + \frac{df(x_2)}{f(x_2)\,ds} + \ldots + \frac{df(x_n)}{f(x_n)\,ds} = 0, \\ \frac{dP}{P\,dt} &= \frac{df(x_1)}{f(x_1)\,dt} + \frac{df(x_2)}{f(x_2)\,dt} + \ldots + \frac{df(x_n)}{f(x_n)\,dt} = 0, \end{aligned}$$

etc. etc.

the first being differentiated with reference to s, the second with reference to t, the last to z, and so on. If in these we place

(5) $\quad df(x_1) = \phi(x_1) f(x_1) dx_1, \quad df(x_2) = \phi(x_2) f(x_2) dx_2,$ etc.

they become

(4)
$$\phi(x_1) \frac{dx_1}{ds} + \phi(x_2) \frac{dx_2}{ds} + \ldots + \phi(x_n) \frac{dx_n}{ds} = 0,$$

$$\phi(x_1) \frac{dx_1}{dt} + \phi(x_2) \frac{dx_2}{dt} + \ldots + \phi(x_n) \frac{dx_n}{dt} = 0,$$

etc.   etc.

and being as many in number as there are unknown quantities they will determine the values of those unknown quantities as soon as we know the form of the function ϕ.

Since these equations are general and applicable to any number of unknown quantities the form of the function ϕ may be determined from any special but known case. Such is that in which there is but one unknown quantity and the observations (8) taken directly upon that quantity. Thus if there be only the quantity s and the measurements give for it the values $M_1, M_2 \ldots M_n$, the errors are

$$x_1 = s - M_1, \quad x_2 = s - M_2 \ldots x_n = s - M_n,$$

from which

$$\frac{dx_1}{ds} = \frac{dx_2}{ds} = \ldots = \frac{dx_n}{ds} = 1,$$

and the first equation in (4) becomes

(6) $\quad \phi(x_1) + \phi(x_2) + \phi(x_3) + \ldots + \phi(x_n) = 0.$

In this case also the arithmetical mean is the most probable value, and the algebraic sum of the errors will be zero (Art. 11a), or

(7) $\quad x_1 + x_2 + x_3 + \ldots + x_n = 0,$

and equations (6) and (7) can only agree when

(8) $\quad \phi(x_1) + \phi(x_2) + \ldots + \phi(x_n) = kx_1 + kx_2 + \ldots + k(x_n),$

in which k is any constant. Replacing from (5) the values of $\phi(x_1)$, $\phi(x_2)$, etc., it becomes

(8) $$\frac{df(x_1)}{f(x_1)\,dx_1} + \frac{df(x_2)}{f(x_2)\,dx_2} + \text{etc.} = kx_1 + kx_2 + \text{etc.};$$

and since this is true whatever be the number of observations, the corresponding terms in the two members are equal. Hence if x be any error and $y = f(x)$, we have

(9) $$\frac{df(x)}{f(x)\,dx} = \frac{dy}{y\,dx} = kx.$$

Multiplying both members by dx and integrating, we obtain

(10) $$\log y = \frac{kx^2}{2} + k',$$

which by passing from logarithms to numbers becomes

(11) $$y = e^{\frac{1}{2}kx^2} e^{k'}.$$

As shown before the constant k must be essentially negative; replacing it then by $-2h^2$, and also placing $e^{k'} = c$, we have

(12) $$y = ce^{-h^2 x^2},$$

as the equation of the probability curve, x being the abscissa, y the ordinate, and c and h constants depending upon the precision of the observations. Considering c and h as unity the values of y corresponding to a few values of x are given in Part I. from which Fig. 1 has been constructed, the vertical scale being double the horizontal in order to exhibit more clearly the form of the curve.

The ordinate of the curve for $x = 0$ is $y = c$, and hence c is the probability of the occurrence of the error x. The probability of committing any *given* error x' is from the equation $y = ce^{-h^2 x^2}$, and this probability will be smaller as h is larger. Hence h is a *measure of precision* of the measurements. The more accurate the observations, the greater is h.

12. For all kinds of observations, then, the law of the probability of error is expressed by the equation

(12) $$y = ce^{-h^2 x^2},$$

132   DEDUCTION OF THE FUNDAMENTAL PRINCIPLES.

which corresponds to the curve of Fig. 1. In discussing the properties and the consequences of the law, it is convenient to know the values of the constants c and h. We proceed to find c (for h, see Arts. 25, 30).

Let x_1, x_2, $x_3 \ldots x_n$ be a series of errors, x_1 being the smallest, x_2 the next following, and x_n the last, the differences between the successive values being equal (thus if the measurement be made by a rule graduated to millimeters, x_2 is 1^{mm} greater than x_1, x_3 is 1^{mm}. greater than x_2, and so on). Then by Art. 7, the probability of committing *one* of these errors, that is, the probability of committing an error lying between x_1 and x_n, is the sum of the separate probabilities $ce^{-h^2 x_1^2}$, $ce^{-h^2 x_2^2}$, etc., or if P' denote this sum

$$(13) \quad P' = c\left(e^{-h^2 x_1^2} + e^{-h^2 x_2^2} + e^{-h^2 x_3^2} + \ldots + e^{-h^2 x_n^2}\right),$$

which may be written

$$(13) \quad P' = c\sum_{x_1}^{x_n} e^{-h^2 x^2},$$

which denotes the sum of the probabilities of all the errors from x_1 to x_n inclusive. P' denotes then the probability that an error lies between the limits x_1 and x_n. Now if i denote the small interval between the successive values of x, and if our observations are accurate enough so that x may be regarded as a continuous variable, i will be very small and equal to dx. Then

$$(14) \quad i\sum_{x_1}^{x_n} e^{-h^2 x^2} = \int_{x_1}^{x_n} e^{-h^2 x^2}\, dx,$$

from which by comparison with (13) we have

$$(15) \quad P' = c\sum_{x_1}^{x_n} e^{-h^2 x^2} = \frac{c}{i}\int_{x_1}^{x_n} e^{-h^2 x^2}\, dx,$$

which expresses the probability that an error will lie between the limits x_1 and x_n. Now it is *certain* that the error will lie between $-\infty$ and $+\infty$, and as unity is the symbol for certainty (Art. 5), we have

$$(16) \quad 1 = \frac{c}{i}\int_{-\infty}^{+\infty} e^{-h^2 x^2}\, dx,$$

DEDUCTION OF THE FUNDAMENTAL PRINCIPLES.   133

from which the value of the constant c will be known as soon as we find the value of the integral between these limits. The integral calculus gives *

$$(17) \qquad \int_{-\infty}^{+\infty} e^{-h^2 x^2}\, dx = \frac{\sqrt{\pi}}{h},$$

* The most convenient method of determining this integral is the geometrical one due to POISSON. I give the process nearly as presented by STURM in his *Cours d'Analyse*, Paris, 1857, Vol. II. p. 16.

Since $y = e^{-h^2 x^2}$ is the equation of a curve of the same form as shown in Fig. 1, the integral $\int y\, dx = \int e^{-h^2 x^2}$ expresses the area between that curve and the axis of x; and since the curve is symmetrical to the axis of Y, that integral between the limits $-\infty$ and $+\infty$ will be equal to double the integral between the limits 0 and $+\infty$. Placing also $hx = t$, we have

$$\int_{-\infty}^{+\infty} e^{-h^2 x^2} dx = \frac{2}{h} \int_0^\infty e^{-t^2} dt,$$

and we have to determine the integral in the second member.

Fig. 15.

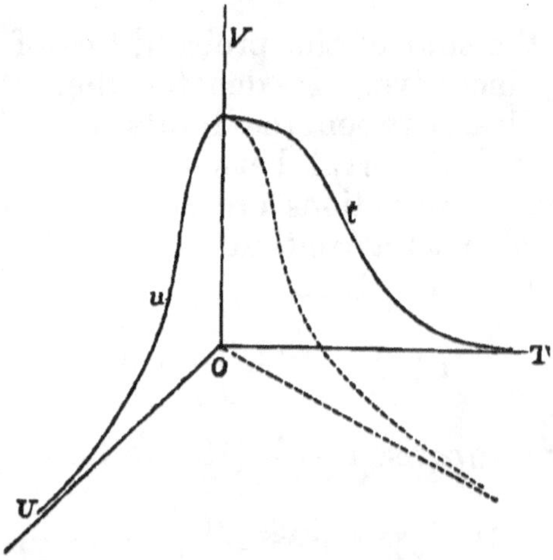

If we take three co-ordinate rectangular axes OT, OU, and OV, and change t into u, we have

$$A = \int_0^\infty e^{-t^2} dt = \text{area between curve } VtT \text{ and axes,}$$

$$A = \int_0^\infty e^{-u^2} du = \text{area between curve } VuU \text{ and axes,}$$

and $\qquad A^2 = \int_0^\infty \int_0^\infty e^{-t^2 - u^2} dt\, du.$

and hence from (16),

(18) $$1 = \frac{c\sqrt{\pi}}{ih} \text{ or } c = \frac{hi}{\sqrt{\pi}};$$

inserting this in (12), we have for the equation of the probability curve, or the law of the probability of error,

(19) $$y = hi\pi^{-\frac{1}{2}}e^{-h^2x^2},$$

in which x is any error, y its probability, h the measure of precision (a quantity so that $\frac{1}{h}$ is a concrete number of the same kind as x), and i is the smallest graduated division in the instrument of measurement, and hence a quantity of the same kind as x or $\frac{1}{h}$. The probability y is then an abstract number, as of course it ought to be.

Inserting also the value of c in equation (15), we have

(20) $$P' = \frac{h}{\sqrt{\pi}} \int_{x_1}^{x_n} e^{-h^2x^2} dx,$$

which expresses the probability that an error will fall between the limits x_1 and x_n. Also since the integral between the

---

Now $v = e^{-t^2}$ is the equation of the curve VtT and $v = e^{-u^2}$ is the equation of VuU, and if either of these curves revolves about the axis of V it generates a surface whose equation is $v = e^{-t^2-u^2}$. Hence the double integral A^2 is one-fourth of the volume included between that surface and the horizontal plane. If we suppose a series of cylinders concentric with the axis V to form the volume, the area of the ring included between two whose radii are r and $r + dr$ is $2\pi r dr$, and the corresponding height is $v = e^{-t^2-u^2} = e^{-r^2}$. Hence one-fourth of the volume is

$$A^2 = \frac{1}{4} \int_0^\infty e^{-r^2} 2\pi r dr,$$

which, since $\int e^{-r^2} 2r dr = -e^{-r^2}$, is equal to $\frac{\pi}{4}$. Therefore we have

$$A = \int_0^\infty e^{-t^2} dt = \frac{\sqrt{\pi}}{2},$$

and hence, finally,

(17) $$\int_{-\infty}^{+\infty} e^{-h^2x^2} dx = \frac{2}{h} \int_0^\infty e^{-t^2} dt = \frac{\sqrt{\pi}}{h}.$$

limits $-x$ and $+x$ is twice the integral from $-x$ to 0 or from 0 to $+x$, we have

$$(21) \qquad P' = \frac{h}{\sqrt{\pi}} \int_{-x}^{+x} e^{-h^2 x^2} dx = \frac{2h}{\sqrt{\pi}} \int_0^x e^{-h^2 x^2} dx,$$

as the probability that an error taken at random is between the limits $-x$ and $+x$, or is numerically less than x.

Now (19) is the equation of the probability curve, and the area between the curve and the axis of x is

$$\int y\, dx = \frac{hi}{\sqrt{\pi}} \int e^{-h^2 x^2} dx.$$

Hence if (16) be multiplied by i it will be the total area of the curve, and if (21) be multiplied by i it will be the area between the limits $-x$ and $+x$. Hence expressions (16), (20) and (21) are *proportional* to the areas of the probability curve corresponding to those limits, and if we regard the total area as unity, the partial area between the limits $-x$ and $+x$ will be a fraction given by P' in equation (21). Further, since errors are committed in proportion to their probabilities, these integrals and their corresponding areas are proportional to the number of errors which we should expect to find between those limits. If then we compute values of P' corresponding to successive numerical values of x in equation (20) they will be fractions proportional to the number of errors numerically less than x, and at the same time express the probabilities of committing an error less than x. As however the constant h depends upon the precision of the measurements, and hence varies in different sets of observations, we write equation (21) under the form

$$(22) \qquad P' = \frac{2}{\sqrt{\pi}} \int_0^{hx} e^{-h^2 x^2} d\,.hx = \frac{2}{\sqrt{\pi}} \int_0^t e^{-t^2} dt,$$

and compute the values of P corresponding to successive numerical values of hx or t, by the usual methods of the integral calculus*. A table of these values is given in Part I.

---

* Developing e^{-t^2} into a series by McLaurin's theorem, multiplying by dt and integrating, we get

13. Thus from our table we see that the probability of an error corresponding to $hx = \infty$ is 1, that is, it is certain that all errors will be less than $x = \frac{\infty}{h}$. For $hx = 1\cdot24$ we have $P' = 0\cdot9205$, that is, the probability that an error will be committed less than $x = \frac{1\cdot24}{h}$ is $\frac{9205}{10000}$. Or in other words, if we have 10000 observations we should expect that in 9205 of them the errors would be less and the remaining 795 greater than $\frac{1\cdot24}{h}$.

The Principle of Least Squares.

14. The law of the probability of error represented by the equation of the probability curve leads directly to important results. The equation is

$$(12) \qquad y = ce^{-h^2x^2},$$

in which y is the probability of committing the error x, and c and h are constants independent of x. Now, considering the general case of independent indirect observations (Art. 19), let n equally good observations be made upon a quantity

$$P' = \frac{2}{\sqrt{\pi}}\left(t - \frac{t^3}{3} + \frac{1}{1.2}\cdot\frac{t^5}{5} - \frac{1}{1.2.3}\cdot\frac{t^7}{7} + \text{etc.}\right),$$

which is convenient for small values of t. For large values we integrate by parts, thus

$$\int e^{-t^2} dt = -\frac{1}{2t}e^{-t^2} - \frac{1}{2}\int \frac{e^{-t^2}}{t}\,dt$$

$$= -\frac{1}{2t}e^{-t^2} + \frac{1}{2^2 t^3}e^{-t^2} + \frac{3}{2^2}\int \frac{e^{-t^2}}{t^4}\,dt = \text{etc.},$$

and since $\int_0^\infty e^{-t^2} dt = \frac{\sqrt{\pi}}{2}$ as shown in the preceding footnote, we have

$$\int_0^t e^{-t^2} dt = \frac{\sqrt{\pi}}{2} - \int_t^\infty e^{-t^2} dt,$$

from which $P' = 1 - \frac{e^{-t^2}}{t\sqrt{\pi}}\left[1 - \frac{1}{2t^2} + \frac{1.3}{(2t^2)^2} - \frac{1.3.5}{(2t^2)^3} + \text{etc.}\right].$

From these two series the values of P' can be found to any required degree of accuracy for all values of t or hx.

DEDUCTION OF THE FUNDAMENTAL PRINCIPLES. 137

$M = f(s, t \ldots z)$ for the purpose of determining the magnitude of $s, t \ldots z$, and let the observed results be $M_1, M_2 \ldots M_n$, each of which is a certain function of $s, t \ldots z$. Subtracting each observed value from the corresponding true value there result the errors $x_1, x_2 \ldots x_n$, having the respective probabilities

(12) $\quad y_1 = ce^{-h^2 x_1^2}, \quad y_2 = ce^{-h^2 x_2^2} \ldots \ldots y_n = ce^{-h^2 x_n^2}.$

Now, by Art. 8, the probability of committing all these errors either simultaneously or successively is the product of the separate probabilities, or

(23) $\quad P = ce^{-h^2 x_1^2} ce^{-h^2 x_2^2} \ldots \ldots ce^{-h^2 x_n^2} = c^n e^{-h^2 \Sigma x^2},$

in which Σx^2 denotes the sum of the squares of the n errors, or

(24) $\quad \Sigma x^2 = x_1^2 + x_2^2 + x_3^2 + \ldots + x_n^2.$

Now each of these errors is a function of the quantities $s, t \ldots z$, which we are to determine. Their *true* values we can never be sure of having obtained however accurate be the measurements. Moreover, in equation (23) the probability P will vary with these quantities, or as Σx^2 takes different values, and hence out of the many systems of values which many be assigned to $s, t \ldots z$, we must take the *most probable* as approaching nearest to the true system, that is the system for which P is a maximum (Art. 9), and P is a maximum when Σx^2 is a *minimum*. Hence *the most probable system of values of observed quantities is that which renders the sum of the squares of the errors a minimum.*

This is the fundamental principle of least squares for observations of equal precision. If they are not equally good, the constants c and h are different in each observation or set of observations, and we have

(12) $\quad y_1 = c_1 e^{-h_1^2 x_1^2}, \quad y_2 = c_2 e^{-h_2^2 x_2^2} \ldots \ldots y_n = c_n e^{-h_n^2 x_n^2}$

as the respective probabilities of the errors $x_1, x_2, \ldots x_n$. Then as before the probability P of the total system of errors is

(25) $\quad P = c_1 c_2 c_3 \ldots c_n e^{-(h_1^2 x_1^2 + h_2^2 x_2^2 + \ldots + h_n^2 x_n^2)},$

and the most probable system is that for which P is a maximum, or that for which

(26) $\quad \Sigma h^2 x^2 = h_1^2 x_1^2 + h_2^2 x_2^2 + \ldots + h_n^2 x_n^2 =$ a minimum.

Hence in measurements of unequal precision, *the square of each error must be multiplied by the square of its measure of precision and the sum of the products be made a minimum.* If h is the same for all errors this evidently reduces to the rule previously given. Thus arises the term "Least Squares."

15. Thus if all the observations are equally good and made directly upon one and the same quantity M whose true value is z, the errors are

$$x_1 = z - M_1, \quad x_2 = z - M_2 \ldots x_n = z - M_n.$$

and the expression (24) is to be made a minimum, or

(24) $\quad \Sigma (z - M)^2 =$ a minimum.

By the usual process for determining minima this gives

$$\Sigma (z - M) = 0, \text{ from which } z = \frac{\Sigma M}{n},$$

which is merely the expression of the law of the arithmetical mean, or of the equation (7), which we took as the foundation of our reasoning.

The Measure of Precision and the Probable Error.

16. Let us consider two series of observations, one having the measure of precision h_1 and the other h_2. The probability that a single error in the first series will lie between the limits $-x_1$ and $+x_1$ will be expressed by the integral

(21) $\quad P_1' = \frac{2h_1}{\sqrt{\pi}} \int_0^{x_1} e^{-h_1^2 x^2} dx,$

and the probability that an error in the second series will fall between the limits $-x_2$ and $+x_2$ is also

(21) $\quad P_2' = \frac{2h_2}{\sqrt{\pi}} \int_0^{x_2} e^{-h_2^2 x^2} dx.$

These two integrals are equal when $h_1 x_1 = h_2 x_2$. If the first series is three times as precise as the second, we have $h_1 = 3h_2$, and hence the integrals will be equal when $x_2 = 3x_1$, that is, the probability of committing an error less than x_2 in the first series is the same as that of committing three times as large an error in the second series. Hence the accuracy of different sets of observations is directly proportional to their measures of precision.

Owing however to the circumstance that h is a quantity of the same kind as $\dfrac{1}{x}$, and hence expressed in terms of an inconvenient unit, it is usual to employ other constants for the comparison of the accuracy of sets of measurements. The one in most common use is called the *probable error*, which is an error of such a value that the probability (22) is $\dfrac{1}{2}$: and is hence an error such that it is an even wager that an error taken at random will be greater or less than it. The probable error is then the value of x given by the equation

(27) $$\frac{1}{2} = \frac{2}{\sqrt{\pi}} \int_0^{hx} e^{-h^2 x^2} d.hx.$$

By interpolation from the table in Part I. Art. 13, we find

for $hx = 0\cdot 4769,\ P' = 0\cdot 5$;

hence denoting this value of x by r, we have for the probable error

(28) $$hr = 0\cdot 4769 \text{ or } r = \frac{0\cdot 4769}{h}.$$

Hence if in Fig. 1 we lay off the abscissæ $-OP$ and $+OP$ equal to the probable error r, and draw the ordinates PB and PB, the area $PBABP$ will be one-half of the total area of the curve, and in any series of observations we may expect that one-half of the errors will be less numerically than r, and the other half greater than r.

17. If then we have two sets of observations whose measures of precision are h_1 and h_2 and probable errors r_1 and r_2, we have

(28) $$h_1 r_1 = 0\cdot 4769,\ h_2 r_2 = 0\cdot 4769,$$

and hence $h_1 r_1 = h_2 r_2$. Now if the precision of the first series is three times that of the second $h_1 = 3h_2$, and hence $r_2 = 3r_1$, that is, the probable error of the first series is one-third that of the second. Hence the probable error serves to compare the accuracy of measurements equally as well as measures of precision. The smaller the probable error, the better are the observations. Thus if two sets of observations give for the length of a line in centimeters

$$L_1 = 427\cdot 32 \pm 0\cdot 04 \text{ and } L_2 = 427\cdot 31 \pm 0\cdot 16,$$

in which 0·04 and 0·16 are the respective probable errors, the meaning is that it is an even wager that the first is within 0·04 of the truth, and also an even wager that the second is within 0·16 of the true value; and the precision of the measurements in the first set is four times that of those in the second.

We have now given the fundamental theory of least squares and probable errors, and shall proceed in the next Chapter to develope its practical features.

CHAPTER VII.

DEVELOPMENT OF PRACTICAL METHODS AND FORMULÆ.

18. WE distinguish the following kinds of observations: *Direct* observations upon a single quantity, in which the measurements are made directly upon that quantity.

19. *Indirect* observations upon one or more quantities $s, t \ldots z$ by the measurement of functions of those quantities (Arts. 11b and 14).

20. *Conditioned* observations, which considered singly are independent, but which collectively are subject to rigorous requirements or conditions. They may be either direct or indirect.

21. *Independent* observations, which are also either direct or indirect, but between which there exists no conditional requirements.

Thus if the sides and angles of a field are measured, each observation *taken alone* is direct. If we find its area from the sides and angles the measurement of that area is indirect. Further, any two sides considered are independent of each other, but if we consider *all* the sides and angles they must fulfil the condition that when plotted they shall form a closed figure.

Direct Observations upon one Quantity.

22. We take up first the case of direct observations *of equal precision* upon one and the same quantity. These are combined by the use of the principle of the average or the arithmetical mean, whose use is limited to this single class.

23. As stated in Art. 11a, the most probable value of a quantity which is measured n times with the results M_1, $M_2 \ldots M_n$ is the arithmetical mean, or

$$(29) \qquad z_0 = \frac{\Sigma M}{n} = \frac{M_1 + M_2 + M_3 + \ldots + M_n}{n}.$$

From the second fundamental axiom of Art. 10, which asserts that in a great number of observations positive and negative errors are equally probable, it would seem that the average was also the true value of the measured quantity. *As far as our observations show,* this is the case, for it is the most accurate value deducible from them, and must be used as if it were the true value. For an infinite number of measurements the average would be the absolute true value; for a limited number it can only be regarded as the most probable value, that is, as the nearest approximation we are able to make to the true value.

24. *Probable errors.* Having taken the average of n equally precise measurements upon a single quantity, we next proceed to investigate the accuracy or precision of the result.

Let the observations give the values M_1, $M_2 \ldots M_n$ whose mean is z_0, and whose corresponding *true* value is z. Let the errors committed in the several observations be

$$x_1, \; x_2 \ldots x_n,$$

so that

$$(30) \qquad x_1 = z - M_1, \qquad x_2 = z - M_2 \ldots x_n = z - M_n.$$

Also let $v_1, v_2 \ldots v_n$ be differences resulting from subtracting each observed value from the arithmetical mean, or

$$(31) \qquad v_1 = z_0 - M_1, \qquad v_2 = z_0 - M_2 \ldots v_n = z_0 - M_n.$$

If z_0 were the true value of the quantity, or $z_0 = z_1$, then the errors x would be the same as the differences v. But as we can never be sure that z_0 represents the true value, we can never determine the errors $x_1, x_2 \ldots x_n$. The values $v_1, v_2 \ldots v_n$, which are readily found in any particular case, we call *residuals;* they should be carefully distinguished from errors.

Each of our measurements $M_1, M_2 \ldots M_n$ is probably incorrect, as likewise the arithmetical mean z_0. The degree of confidence which we can place in each of these results will be shown by their probable errors (Art. 16). Considering h as the measure of precision of a single observation and r as its probable error, the relation between them is given by

(28) $$r = \frac{0.4769}{h},$$

and hence to determine r we have only to find h. There is, however, no known method of finding the *exact* value of h: the best that we can do is to determine an approximate value, which moreover shall be the *most probable* one (Art. 9).

25*. The probability of the occurrence of any error x is

(19) $$y = h i \pi^{-\frac{1}{2}} e^{-h^2 x^2},$$

and the probability of the occurrence of the system of errors $x_1, x_2 \ldots x_n$ is by Art. 8 the product of the separate probabilities. If then h is the same for each of the n observations, we have as in Art. 14,

(32) $$P = y_1 y_2 \ldots y_n = h^n i^n \pi^{-\frac{1}{2}n} e^{-h^2 \Sigma x^2},$$

where Σx^2 denotes the sum

$$x_1^2 + x_2^2 + x_3^2 + \ldots + x_n^2.$$

Now in this expression h is unknown, and further, we have no means of finding its exact value. But whatever be its value, P must be a maximum in order to give the most probable value of the measured quantity z. We are therefore led to conclude that for a *given* system of errors the most probable value of h is that which renders P a maximum. Differentiating (32) with reference to h, and putting the first differential coefficient equal to zero, we have

(33) $$\frac{dP}{dh} = n h^{n-1} i^n \pi^{-\frac{1}{2}n} e^{-h^2 \Sigma x^2} - 2 h \Sigma x^2 e^{-h^2 \Sigma x^2} h^n i^n \pi^{-\frac{1}{2}n} = 0;$$

---

* Simplified and considerably altered from the demonstration given by DIENGER in his *Ausgleichung der Beobactungsfehler*, Braunschweig, 1857, p. 59.

dividing this equation by h^{n-1}, i^n, $\pi^{-\frac{1}{2}n}$ and $e^{-h^2\Sigma x^2}$, we obtain

(33) $$n - 2h^2\Sigma x^2 = 0,$$

from which

(34) $$\Sigma x^2 = \frac{n}{2h^2} \text{ and } h^2 = \frac{n}{2\Sigma x^2}.$$

We have thus the value of h in terms of Σx^2, which, however, we have no means of obtaining, since the errors $x_1, x_2 \ldots x_n$ depend upon the unknown true value z. If the number of observations were infinite Σx^2 would equal Σv^2 (Art. 23), and as the latter value is determinate h would be known. In a large number of observations, therefore, the equation

$$h^2 = \frac{n}{2\Sigma v^2}$$

will always give a close approximation to the value of h. But as the sum Σv^2 is always less than Σx^2 (since from the principle of least squares, Art. 14, the first is the minimum value of the second) we may place

(35) $$\Sigma x^2 = \Sigma v^2 + k^2,$$

in which k^2 is a constant to be determined, and then our value of h will be correctly given by inserting in (34) the value of Σx^2 from (35). As however Σx^2 cannot be exactly found, we cannot hope to find the exact value of k^2, but must be content with determining an approximate one.

Now the probability of committing the system of errors $x_1, x_2 \ldots x_n$ is

(32) $$P = c^n e^{-h^2\Sigma x^2},$$

or, inserting for Σx^2 its value from (35),

(36) $$P = c^n e^{-h^2(\Sigma v^2 + k^2)} = c^n e^{-h^2\Sigma v^2} e^{-h^2k^2}.$$

Placing in this the constant terms equal to c, it is

(36) $$P = c e^{-h^2k^2}.$$

Hence the law of the probability of any value k is the same as that of an error x, as shown by (12). We may regard then (36) as the equation of a curve of the same form as Fig. 1, and, as in Art. 12, we may show that $c = hi\pi^{-\frac{1}{2}}$, where i is a small constant of the same kind as k. Hence the equation

(37) $$P = hi\pi^{-\frac{1}{2}} e^{-h^2 k^2}$$

shows the probability P of a value k. Now in this equation both h and k are unknown, and, as we have said before, we can only expect to determine their most probable values. The value of k^2 although unknown is fixed and definite, and hence we conclude as before that the most probable value of h will be that which makes P a maximum (Art. 9). Differentiating then the equation (37) with reference to h and placing the first differential coefficient equal to zero, we have

(38) $$\frac{dP}{dh} = i\pi^{-\frac{1}{2}} e^{-h^2 k^2} - 2hk^2 e^{-h^2 k^2} hi\pi^{-\frac{1}{2}} = 0.$$

Dividing this by i, $\pi^{-\frac{1}{2}}$ and $e^{-h^2 k^2}$, we have

(38) $$1 - 2h^2 k^2 = 0,$$

from which

(39) $$k^2 = \frac{1}{2h^2}.$$

Therefore for the equation (35), we have

(40) $$\Sigma x^2 = \Sigma v^2 + \frac{1}{2h^2}$$

as the nearest possible approximation. Since from (33) the value of Σx^2 is $\frac{n}{2h^2}$, we have accordingly,

(40) $$\frac{n}{2h^2} = \Sigma v^2 + \frac{1}{2h^2},$$

from which we find the most probable value of h in terms of the known sum of the squares of the residuals (31), or

(41) $$h = \sqrt{\frac{n-1}{2\Sigma v^2}}.$$

Inserting this value in (29) we have as the probable error of a single observation

$$(42) \qquad r = 0{\cdot}4769 \sqrt{\frac{2\Sigma v^2}{n-1}} = 0{\cdot}6745 \sqrt{\frac{\Sigma v^2}{n-1}}.$$

The above is only one of the many demonstrations of the formula for the probable error r. None of them are, from the nature of the case, entirely satisfactory, since it is impossible to find the exact value of h. In Art. 30 we give another demonstration which may be readily applied to the case here considered by making $g = 1$, or regarding the measurements as of equal precision.

26. The formula (42) just deduced gives the probable error of a single measurement M. We next inquire, what is the probable error of the arithmetical mean of the n measurements?

The probability of the arithmetical mean is the probability of committing the system of errors $v_1, v_2 \ldots v_n$ or

$$(23) \qquad P_0 = c^n e^{-h^2 \Sigma v^2},$$

and the probability that the true value of z is $z_0 + x_1'$ is the probability of the system of errors $v_1 + x'$, $v_2 + x'$, etc., or

$$(43) \qquad P_{x'} = c^n e^{-h^2 \Sigma (v+x')^2}.$$

Since $\Sigma (v + x')^2 = \Sigma v^2 + 2x' \Sigma v + n x'^2$ and $\Sigma v = 0$, this becomes

$$(43) \qquad P_{x'} = c^n e^{-h^2 (\Sigma v^2 + n x'^2)}.$$

Hence we have

$$(44) \qquad P_0 : P_{x'} :: e^{-h^2 \Sigma v^2} : e^{-h^2 (\Sigma v^2 + n x'^2)} :: 1 : e^{-n h^2 x'^2},$$

that is, the probability of the error 0 in the arithmetical mean is to that of the error x' as 1 is to $e^{-n h^2 x'^2}$. For a single observation whose error is x', we have however from (2)

$$(45) \qquad y_0 : y_x :: 1 : e^{-h^2 x'^2},$$

or the probability of the error 0 in a single observation is to that of the error x' as 1 is to $e^{-h^2 x'^2}$. Hence comparing (44) and (45) we see that if h is the measure of precision of a single observation, $\sqrt{nh}$ must be the measure of precision of the arithmetical mean z_0. Therefore, h_0 denoting the measure of precision of z_0, we have

(46) $$h_0 = h\sqrt{n}.$$

Denoting by r_0 the probable error of z_0,

(28) $$h_0 r_0 = 0{\cdot}4769 \text{ and } hr = 0{\cdot}4769.$$

Inserting from these the values of h_0 and h in (46) we have $r_0 = \dfrac{r}{\sqrt{n}}$, that is, *the probable error of the arithmetical mean is equal to the probable error of a single observation divided by the square root of the number of observations.* Hence from (42)

(47) $$r_0 = \frac{r}{\sqrt{n}} = 0{\cdot}6745\sqrt{\frac{\Sigma v^2}{n(n-1)}},$$

which is the expression used without proof in Part I.

27. We take up next the adjustment of direct observations of *unequal precision*. If the several observations give the results $M_1, M_2 \ldots M_n$ whose measures of precision are $h_1, h_2 \ldots h_n$, the principle of least squares requires that the quantity

(26) $$h_1^2 x_1^2 + h_2^2 x_2^2 + \ldots + h_n^2 x_n^2 = \text{a minimum}.$$

If z be the true value of the quantity the errors are

$$x_1 = z - M_1, \quad x_2 = z - M_2 \ldots x_n = z - M_n,$$

and the expression to be made a minimum is

(26) $$h_1^2(z - M_1)^2 + h_2^2(z - M_2)^2 + \ldots + h_n^2(z - M_n)^2.$$

By differentiation, we find that the value of z which makes this a minimum is

(48) $$Z = \frac{h_1^2 M_1 + h_2^2 M_2 + \ldots + h_n^2 M_n}{h_1^2 + h_2^2 + \ldots + h_n^2} = \frac{\Sigma h^2 M}{\Sigma h^2}.$$

Owing however to the fact already alluded to in Art. 16, that h is expressed in terms of an inconvenient unit it is usual to employ numbers having the same ratios; thus if

(49) $$g_1 : g_2 : g_n :: h_1^2 : h_2^2 : h_n^2,$$

the expression (48) becomes

(50) $$Z = \frac{g_1 M_1 + g_2 M_2 + \cdots + g_n M_n}{g_1 + g_2 + \cdots + g_n} = \frac{\Sigma g M}{\Sigma g}.$$

The numbers $g_1, g_2 \ldots g_n$ are called the *weights* of the observations $M_1, M_2 \ldots M_n$, and are merely relative numbers proportional to the absolute quantities $h_1^2, h_2^2 \ldots h_n^2$. If, as in Art. 23, all the observations are of equal precision, we take g as 1, and the formula (50) agrees with the law of the arithmetical mean (27). The weights of measurements are then numbers proportional to the number of single observations to which each is equivalent. Thus if M_1 is the equivalent of g_1 single observations of the weight unity its weight is g_1, and as the arithmetical mean z_0 is the equivalent of n single observations its weight is n. The weight of Z, which we call the *general mean*, is then $g_1 + g_2 + \cdots + g_n$.

28. If we have different sets of measurements of unequal precision upon one quantity, the first giving the average z_1 from n_1 measurements, the second giving z_2 from n_2 measurements, the adjusted value is furnished by

(50) $$Z = \frac{\Sigma g z}{\Sigma g},$$

and the weights $g_1, g_2 \ldots g_n$ must be found from the proportion (49). The measures of precision of the several averages being h_1, h_2, etc., we have from equations (41) and (46)

(51) $$h_1^2 = \frac{n_1(n_1 - 1)}{2\Sigma v'^2}, \quad h_2^2 = \frac{n_1(n_1 - 1)}{2\Sigma v''^2},$$

in which $\Sigma v'^2$ denotes the sum of the squares of the residuals in the first set, $\Sigma v''^2$ in the second, and so on. Inserting these in (49) and omitting the common factor 2, we have

(52) $$g_1 : g_2 : g_3 :: \frac{n_1(n_1 - 1)}{\Sigma v'^2} : \frac{n_2(n_2 - 1)}{\Sigma v''^2} : \frac{n_3(n_3 - 1)}{\Sigma v'''^2},$$

from which we may find the relative weights, and then the general mean by (50). The weight of the general mean is of course Σg.

29. If the probable errors of the averages of different sets of measurements have been found by (47), their relative weights are easily determined by (49). For if the measures of precision be $h_1, h_2 \ldots h_n$, and the corresponding probable errors be $r_1, r_2 \ldots r_n$, we have

$$(28) \qquad h_1^2 = \frac{0.4769}{r_1}, \quad h_2 = \frac{0.4769}{r_2}, \text{ etc.;}$$

inserting these in (49) and omitting the common factor 0·4769, we have

$$(53) \qquad g_1 : g_2 : g_3 :: \frac{1}{r_1^2} : \frac{1}{r_2^2} : \frac{1}{r_3^2},$$

that is, the weights of observations are inversely proportional to the squares of their probable errors.

30. We now proceed to find the *probable error* of the general mean. Let Z be the general mean, G its weight, and R its probable error, also let r_1 and r_2 be the probable errors of observations of the weights g_1, g_2. Then from the above principle

$$(53) \qquad G : g_1 : g_2 :: \frac{1}{R^2} : \frac{1}{r_1^2} : \frac{1}{r_2^2},$$

from which we find

$$(54) \qquad R^2 = \frac{g_1 r_1^2}{G} = \frac{g_2 r_2^2}{G} = \text{etc.}$$

Also since $G = g_1 + g_2 + \ldots + g_n$, we have from (53)

$$(55) \qquad \frac{1}{R^2} = \frac{1}{r_1^2} + \frac{1}{r_2^2} + \ldots + \frac{1}{r_n^2}.$$

Hence, having by (46) found the probable errors $r_1, r_2 \ldots r_n$, the probable error of Z may be found by (55), or, having by (52) found the weights $g_1, g_2 \ldots g_n$, whose sum is G, we may by (54) also find the probable error of Z.

As however the computation of all the weights or all the probable errors is sometimes laborious, we shall deduce another formula. Let, as before, G be the weight of the general mean, and R its probable error, and let r be the probable error of an observation whose weight is unity. Then we have

(55) $$G : 1 :: \frac{1}{R^2} : \frac{1}{r^2},$$

from which

(56) $$R = \frac{r}{\sqrt{G}}.$$

Hence *having found the probable error r of an observation whose weight is unity, the probable error of a result whose weight is* G *is found by dividing* r *by the square root of* G. We proceed to develop a method for finding r.

Let n be the number of observations or sets of measurements, $h_1, h_2 \ldots h_n$ their measures of precision, and $g_1, g_2 \ldots g_n$ their relative weights. Also let h be the measure of precision of an observation whose weight is unity. Then from (49)

$$g_1 : g_2 : 1 :: h_1^2 : h_2^2 : h^2,$$

and hence

(57) $$h_1^2 = g_1 h^2, \quad h_2^2 = g_2 h^2 \ldots h_n^2 = g_n h^2.$$

The law of the probability of error gives for the corresponding errors $x_1, x_2 \ldots x_n$,

(19) $$y_1 = h_1 i_1 \pi^{-\frac{1}{2}} e^{-h_1^2 x_1^2}, \quad y_2 = h_2 i_2 \pi^{-\frac{1}{2}} e^{-h_2^2 x_2^2} \text{ etc.,}$$

or from (57),

$$y_1 = h\sqrt{g_1}\, i_1 \pi^{-\frac{1}{2}} e^{-h^2 g_1 x_1^2}, \quad y_2 = h\sqrt{g_2}\, i_2 \pi^{-\frac{1}{2}} e^{-h^2 g_2 x_2^2}.$$

Hence in general if x be *any* error, g the weight of its corresponding observation, and h the measure of precision of an observation *of the weight unity*, we have

(58) $$y = h\sqrt{g}\, i\, \pi^{-\frac{1}{2}} e^{-h^2 g x^2},$$

as the probability y of any error x, whose measure of precision is $h\sqrt{g}$, or as the probability of any error $x\sqrt{g}$ whose measure of precision is h.

Now n observations being made, it is evident that the quantity

$$(59) \quad \frac{g_1 x_1^2 + g_2 x_2^2 + \ldots + g_n x_n^2}{n} = \frac{\Sigma g x^2}{n}$$

will have a certain definite value; the probability of the occurrence of $g_1 x_1^2$ will be y_1, of $g_2 x_2^2$ will be y_2, etc.: and each term in the numerator will occur a number of times proportional to its probability, provided that n is a very large number. Hence $g_1 x_1^2$ occurs $n y_1$ times, $g_2 x_2^2$ occurs $n y_2$ times, and the quantity (59) becomes

$$(59) \quad g_1 x_1^2 y_1 + g_2 x_2^2 y_2 + \ldots + g_n x_n^2 y_n = \Sigma g x^2 y.$$

But when n is a very large number, the errors will be distributed according to the law of the probability curve from $-\infty$ to $+\infty$, and if the measurements are accurate, i in (58) will be dx. Hence we have, by inserting for y its value,

$$(60) \quad \frac{\Sigma g x^2}{n} = \Sigma g x^2 y = \frac{h}{\sqrt{\pi}} \sqrt{g} \int_{-\infty}^{+\infty} g x^2 e^{-h^2 g x^2} dx.$$

Taking in this $hx\sqrt{g} = t$ as the unit variable, it may be written

$$(60) \quad \frac{\Sigma g x^2}{n} = \frac{1}{h^2 \sqrt{\pi}} \int_{-\infty}^{+\infty} t^2 e^{-t^2} dt,$$

and as the value of the integral is $\frac{\sqrt{\pi}}{2}$* we have

---

* From the foot-note to equation (17) we have

$$\int_0^\infty e^{-t^2} dt = \frac{\sqrt{\pi}}{2}.$$

Placing $t = t\sqrt{s}$, this becomes

$$\int_0^\infty e^{-t^2 s} dt = \frac{\sqrt{\pi}}{2\sqrt{s}}.$$

Differentiating this equation with reference to s, and regarding t as constant, we have

$$-\int_0^\infty e^{-t^2 s} t^2 ds\, dt = -\frac{\sqrt{\pi}}{4} \frac{ds}{\sqrt{s^3}}.$$

(61) $$\frac{\Sigma g x^2}{n} = \frac{1}{2h^2},$$

which gives the value of h in terms of the sum $\Sigma g x^2$.

Now let $v_1, v_2 \ldots v_n$ denote the residuals or differences between the general mean Z and each observation. Then the sum $\Sigma g x^2$ is greater than $\Sigma g v^2$, since the second is the minimum of the first. If we place then

(62) $$\Sigma g x^2 - \Sigma g v^2 = k^2,$$

and suppose $\Sigma g x^2$ to have all possible values greater than $\Sigma g v^2$, and each to be repeated a number of times proportional to its probability, we may consider the mean of all the values thus found for k^2 as the best approximation attainable to its value. The law of the probability of these values of k is as in (36) given by the equation

$$Y = c e^{-h^2 k^2},$$

and if n be the number of possible values of k_1^2 the value k_1^2 will occur $n Y_1$ times, k_2^2 will occur $n Y_2$ times, and hence the mean of all the possible values will be

(63) $$\Sigma k^2 Y = \frac{h}{\sqrt{\pi}} \int_{-\infty}^{+\infty} k^2 e^{-h^2 k^2} \, dk = \frac{1}{2h^2}.$$

If then we place in (63) for $\Sigma g x^2$ its value from (62) and for k^2 the value just found, we have

(63) $$\frac{n}{2h^2} = \Sigma g v^2 + \frac{1}{2h^2},$$

from which we find for h

(64) $$h = \sqrt{\frac{n-1}{2\Sigma g v^2}}.$$

Hence from (29) we have the probable error of an observation of the weight unity

(65) $$r = \frac{0 \cdot 4769}{h} = 0 \cdot 6745 \sqrt{\frac{\Sigma g v^2}{n-1}},$$

Dividing this by $-ds$ and making $s = 1$, we obtain

$$\int_0^\infty e^{-t^2} t^2 \, dt = \frac{\sqrt{\pi}}{4} = \text{one-half of the integral in (60).}$$

which being computed, the probable error of the general mean is found at once by (56).

If in (64) and (65) we place $g=1$, they reduce to the expressions (41) and (42) obtained in Art. 25 by an entirely different method. Moreover in that case G becomes n, and the expression (56) coincides with (47). Both methods of reasoning thus lead to the same results.

31. The above includes the whole theory of direct measurements upon a single quantity. Observations upon several quantities will be investigated in the next section; and it is interesting to observe that all the preceding methods are but particular cases of the more general theory of indirect observations, as we shall show in the following articles.

Independent observations upon several quantities.

32. Independent observations are those which are subject to no conditions except those imposed upon them by the measurements themselves, so that, before taking the observations, all systems of values are equally probable. The manner of measurement may be either direct (Art. 18) or indirect (Art. 19), but for convenience we shall consider only the latter, of which the first is a special case.

Indirect observations being made upon functions of the quantities to be determined, require in general the statement of equations between the measured quantities and those required. Thus if in order to determine the magnitude of the quantities $s, t \ldots z$ we make observations upon the related quantities M, M', which are connected with the first by the relations

$$M = f(s, t \ldots z),$$
$$M' = f'(s, t \ldots z),$$

and find the values M_1, M_2, M_1', M_2', etc., each observation furnishes us with an equation, which we call an *observation equation*. The number of these equations is the same as the number of observations, and generally greater than the number of unknown quantities $s, t \ldots z$ which we are to determine.

Hence in general no system of values can be found for $s, t \ldots z$ which will exactly satisfy the observation equations. They may however be approximately satisfied by many sytems of values, and we propose then the problem to find out of those systems, all equally possible, the one which is the *most probable*, and hence the best.

The equations between the observed and unknown quantities may be either linear or non-linear, exponential or transcendental; but we shall treat only of linear equations, to which all the others can always be reduced by the methods of Art. 59.

33. Taking up first the case of *observations of equal weight*, let the equations between the observed and the measured quantities be of the form

$$as + bt + \ldots + lz = M,$$

in which $s, t \ldots z$ are the unknown quantities to be determined, $a, b \ldots l$, constants given by theory and absolutely known, and M the measured quantity. For each observation we shall have a similar equation, and in all the following n equations

(66)
$$a_1 s + b_1 t + \ldots + l_1 z = M_1,$$
$$a_2 s + b_2 t + \ldots + l_2 z = M_2,$$
$$a_3 s + b_3 t + \ldots + l_3 z = M_3,$$
$$\ldots\ldots\ldots\ldots\ldots\ldots\ldots\ldots$$
$$a_n s + b_n t + \ldots + l_n z = M_n,$$

the first of which arises from the first observation, the second from the second, and the last from the nth.

Now as the number of these observation equations is greater than that of the unknown quantities, they will not be exactly satisfied for any system of values we may find. Hence if $s, t \ldots z$ denote the *true* values of the quantities to be determined, the general form

$$as + bt + \ldots + lz - M = 0$$

is only approximately correct, for the equations do not reduce exactly to zero. Let us designate then by $x_1, x_2 \ldots x_n,$

the errors which thus arise when for $s, t \ldots z$ in (66) we place their true values. Then we have strictly

(67)
$$a_1 s + b_1 t + \ldots + l_1 z - M_1 = x_1,$$
$$a_2 s + b_2 t + \ldots + l_2 z - M_2 = x_2,$$
$$\ldots\ldots\ldots\ldots\ldots\ldots\ldots\ldots\ldots\ldots$$
$$a_n s + b_n t + \ldots + l_n z - M_n = x_n.$$

Now in the impossibility of finding the true values of $s, t \ldots z$ from these equations, we must determine their *most probable* values as the nearest attainable approximation to the truth. The most probable system of values is, by the fundamental principle of Art. 14, that which makes the sum of the squares of the errors a minimum, that is which makes $x_1^2 + x_2^2 + \ldots + x_n^2$ a minimum. By the use of this principle we have in Part I. deduced a method of finding the most probable values. We give here a more general proof which follows directly from our demonstration of Art. 11b.

From equations (8) and (9) of that Article we have

$$\phi(x_1) = kx_1, \quad \phi(x_2) = kx_2 \ldots \phi(x_n) = kx_n,$$

in which k is any constant. Substituting these in the differential equations (4) and dividing each by k, we obtain

(68)
$$x_1 \frac{dx_1}{ds} + x_2 \frac{dx_2}{ds} + x_3 \frac{dx_3}{ds} + \ldots + x_n \frac{dx_n}{ds} = 0,$$
$$x_1 \frac{dx_1}{dt} + x_2 \frac{dx_2}{dt} + x_3 \frac{dx_3}{dt} + \ldots + x_n \frac{dx_n}{dt} = 0,$$
$$\ldots\ldots\ldots\ldots\ldots\ldots\ldots\ldots\ldots\ldots\ldots\ldots\ldots$$
$$x_1 \frac{dx_1}{dz} + x_2 \frac{dx_2}{dz} + x_3 \frac{dx_3}{dz} + \ldots + x_n \frac{dx_n}{dz} = 0.$$

Now by differentiating equations (66) with reference to each variable, we obtain

(69)
$$\frac{dx_1}{ds} = a_1, \quad \frac{dx_2}{ds} = a_2 \ldots \frac{dx_n}{ds} = a_n,$$
$$\frac{dx_1}{dt} = b_1, \quad \frac{dx_2}{dt} = b_2 \ldots \frac{dx_n}{dt} = b_n,$$
$$\ldots\ldots\ldots\ldots\ldots\ldots\ldots\ldots\ldots\ldots\ldots$$
$$\frac{dx_1}{dz} = l_1, \quad \frac{dx_2}{dz} = l_2 \ldots \frac{dx_n}{dz} = l_n,$$

which substituted in (68) give

$$a_1 x_1 + a_2 x_2 + a_3 x_3 + \ldots + a_n x_n = 0,$$
$$b_1 x_1 + b_2 x_2 + b_3 x_3 + \ldots + b_n x_n = 0,$$
(70)
$$\ldots\ldots\ldots\ldots\ldots\ldots\ldots\ldots\ldots\ldots\ldots\ldots\ldots$$
$$l_1 x_1 + l_2 x_2 + l_3 x_3 + \ldots + l_n x_n = 0,$$

which are the conditions for determining the most probable values of $s, t \ldots z$, since they render equation (2) a minimum, and which are as many in number as the number of those unknown quantities. If in these we substitute for $x_1, x_2 \ldots x_n$ their values from (67), we have then the equations from which $s, t \ldots z$ can be determined. These final equations we call *normal equations*, and we see by (70) that the first is formed by multiplying each observation equation by the coefficient of s in that equation and adding the results. Inserting in the first of (70) the values of $x_1 x_2 \ldots x_n$ from (67), we have as the first normal equation, or the equation for s,

(71) $(a_1^2 + a_2^2 + \ldots + a_n^2) s + (a_1 b_1 + a_2 b_2 + \ldots + a_n b_n) t + \ldots$
$$+ (a_1 l_1 + a_2 l_2 + \ldots + a_n l_n) z$$
$$- (a_1 M_1 + a_2 M_2 + \ldots + a_n M_n) = 0;$$

and in like manner we form a normal equation for each of the other quantities $t, u \ldots z$. To abbreviate the expression of these equations let us place

$$\Sigma a^2 = a_1^2 + a_2^2 + a_3^2 + \ldots + a_n^2,$$
$$\Sigma ab = a_1 b_1 + a_2 b_2 + a_3 b_3 + \ldots + a_n b_n,$$
(72) $\quad \Sigma al = a_1 l_1 + a_2 l_2 + a_3 l_3 + \ldots + a_n l_n,$
$$\Sigma b^2 = b_1^2 + b_2^2 + b_3^2 + \ldots + b_n^2,$$
$$\Sigma aM = a_1 M_1 + a_2 M_2 + a_3 M_3 + \ldots + a_n M_n,$$

etc. etc. etc.

and then the normal equations (71) may be written

$$(73)\quad \begin{aligned}\Sigma a^2.s + \Sigma ab.t + \ldots + \Sigma al.z &= \Sigma aM,\\ \Sigma ab.s + \Sigma b^2.t + \ldots + \Sigma bl.z &= \Sigma bM,\\ \Sigma ac.s + \Sigma bc.t + \ldots + \Sigma cl.z &= \Sigma cM,\\ &\cdots\cdots\cdots\\ \Sigma al.s + \Sigma bl.t + \ldots + \Sigma l^2.z &= \Sigma lM.\end{aligned}$$

The coefficients of the unknown quantities, it will be remarked, present a curious symmetry; thus the coefficients of the first horizontal row are the same as those of the first vertical row, those of the second horizontal row the same as those of the second vertical row, and so on. The period placed between each coefficient and its unknown quantity shows that the sign Σ extends only to the former.

34. Thus if we have n observations for determining q unknown quantities, the most probable values of the unknown quantities are obtained by writing n observation equations as in (66), then forming the q normal equations as in (73), whose coefficients Σa^2, Σab, etc. are given by (72); then the solution of these normal equations will furnish the most probable values of $s, t \ldots z$. In the most common cases the coefficients in the observation equations (66) are $+1, -1$ or 0, and in the formation of the sums (71) the signs must be carefully regarded. Stated in words, the process for forming the above normal equations is the same as given by the rule deduced in Part I.

35. To adjust indirect independent observations of equal weight, we have then only to form and solve the normal equations, thus obtaining the best system of values for the unknown quantities.

The solution of the normal equations may be effected by any algebraic method. When there are only two or three equations the usual methods of substitution or addition are perhaps the quickest, but for many equations they are tedious. A process of solution by substitution, assisted by the notation of (71), which is due to GAUSS, is here valuable to the computer (see Art. 60). The method of *indeterminate multipliers* is likewise often of quick application. As this method is not presented as fully as it ought to be in the common text-books in Algebra, we think it worth while to give an example illustrating its use, particularly as we shall have occasion to refer to it again in another connection.

We take the five normal equations given in Part I. Art. 35, viz.

$$2s - t = A,$$
$$-s + 4t - u - x = B,$$
$$-t + 2u - x = C,$$
$$-t - u + 3x - y = D,$$
$$-x + 3y = E,$$

and to solve them by the method of indeterminate multipliers, we multiply the first by a number β_1, which is as yet unknown, also the second by a number β_2, the third by β_3, the fourth by β_4, and the fifth by β_5. We then add the resulting equations and place together the terms containing like unknown quantities, thus

$$(2\beta_1 - \beta_2)s + (-\beta_1 + 4\beta_2 - \beta_3 - \beta_4)t + (-\beta_2 + 2\beta_3 - \beta_4)u$$
$$+ (-\beta_2 - \beta_3 + 3\beta_4 - \beta_5)x$$
$$+ (-\beta_4 + 3\beta_5)y = \beta_1 A + \beta_2 B + \beta_3 C + \beta_4 D + \beta_5 E.$$

Now if we wish to find the value of y we have only to require such relations to exist between these multipliers that all of the terms of the first member shall disappear except that containing y, that is, we must have

$$2\beta_1 - \beta_2 = 0,$$
$$-\beta_1 + 4\beta_2 - \beta_3 - \beta_4 = 0,$$
$$-\beta_2 + 2\beta_3 - \beta_4 = 0,$$
$$-\beta_2 - \beta_3 + 3\beta_4 - \beta_5 = 0,$$

and then we find

$$y = \frac{\beta_1 A + \beta_2 B + \beta_3 C + \beta_4 D + \beta_5 E}{-\beta_4 + 3\beta_5}.$$

Now β_1, β_2, etc. may have any values which will satisfy the four imposed conditions. If then we take $\beta_1 = 1$, the first of these conditions gives $\beta_2 = 2$, and the other three become

$$7 - \beta_3 - \beta_4 = 0,$$
$$-2 + 2\beta_3 - \beta_4 = 0,$$
$$-2 - \beta_3 + 3\beta_4 - \beta_5 = 0.$$

Solving these equations by the common algebraic method of addition we find $\beta_3 = 3$, $\beta_4 = 4$, and $\beta_5 = 7$. Hence our value of y is

$$y = \frac{A + 2B + 3C + 4D + 7E}{17}.$$

In like manner if we wish to find the value of s, we require that all the terms shall disappear except the first, or that

$$-\beta_1 + 4\beta_2 - \beta_3 - \beta_4 = 0,$$
$$-\beta_2 + 2\beta_3 - \beta_4 = 0,$$
$$-\beta_2 - \beta_3 + 3\beta_4 - \beta_5 = 0,$$
$$-\beta_4 + 3\beta_5 = 0,$$

and then we have

$$s = \frac{\beta_1 A + \beta_2 B + \beta_3 C + \beta_4 D + \beta_5 E}{2\beta_1 - \beta_2},$$

and as the multipliers may have any values which will satisfy the four conditions, we take $\beta_5 = 1$ from which $\beta_4 = 3$, and hence from the other three $\beta_3 = \frac{11}{3}$, $\beta_2 = \frac{13}{3}$, and $\beta_1 = \frac{32}{3}$. Then

$$s = \frac{32A + 13B + 11C + 9D + 3E}{51}.$$

Thus by the operation of indeterminate multipliers we reduce in this case the five given equations to three much simpler equations whose solution is readily effected by the common methods. The values of the other unknown quantities can now be either found directly from the normal equations by inserting for s and y their values, or by imposing new conditions and finding new sets of values for the indeterminate multipliers.

36. *Observations of unequal weight* next claim our attention. As before, let the observations be represented by the equations (66), the measure of precision of the first being h_1, of the second h_2, etc. and their corresponding weights g_1, g_2, etc. Then the errors being $x_1, x_2 \ldots x_n$, we may as in (67) write them

$$(74)\quad\begin{aligned}a_1 s+b_1 t+\ldots+l_1 z-M_1&=x_1 \text{ with weight } g_1,\\ a_2 s+b_2 t+\ldots+l_2 z-M_2&=x_2 \quad\text{,,}\quad\text{,,}\quad g_2\\ &\cdots\\ a_n s+b_n t+\ldots+l_n z-M_n&=x_n \text{ with weight } g_n.\end{aligned}$$

Now by the principle of least squares (Art. 16), the most probable values of the unknown quantities s, t, z are those that make

$$(26)\quad h_1^2 x_1^2 + h_2^2 x_2^2 + \ldots + h_n^2 x_n^2 = \text{a minimum.}$$

But if in this we place

$$(57)\quad h_1^2 = h^2 g_1,\quad h_2^2 = h^2 g_2 \ldots h_n^2 = h^2 g_n,$$

in which h is the measure of precision of an observation whose weight is unity, it becomes

$$(75)\quad h^2(g_1 x_1^2 + g_2 x_2^2 + g_3 x_3^2 + \ldots + g_n x_n^2) = \text{a minimum,}$$

so that we have to render a minimum the quantity

$$(76)\quad \Sigma g x^2 = g_1 x_1^2 + g_2 x_2^2 + \ldots + g_n x_n^2.$$

Remembering that x_1^2, x_2^2 are functions of the variables $s, t \ldots z$ as given by equations (74), we must to determine the minimum differentiate $\Sigma g x^2$ with reference to each of those variables and place the several differential coefficients equal to zero; thus, after dividing by 2,

$$(77)\quad\begin{aligned}g_1 x_1 \frac{dx_1}{ds} + g_2 x_2 \frac{dx_2}{ds} + \ldots + g_n x_n \frac{dx_n}{ds} &= 0,\\ g_1 x_1 \frac{dx_1}{dt} + g_2 x_2 \frac{dx_2}{dt} + \ldots + g_n x_n \frac{dx_n}{dt} &= 0,\\ &\cdots\\ g_1 x_1 \frac{dx_1}{dz} + g_2 x_2 \frac{dx_2}{dz} + \ldots + g_n x_n \frac{dx_n}{dz} &= 0.\end{aligned}$$

Next differentiating (74) with reference to each variable, we have the coefficients $\frac{dx_1}{ds} = a_1$, $\frac{dx_2}{ds} = a_2$, etc. exactly as in (69), and, inserting these in (77), we have the conditions

(78)
$$g_1 a_1 x_1 + g_2 a_2 x_2 + \ldots + g_n a_n x_n = 0,$$
$$g_1 b_1 x_1 + g_2 b_2 x_2 + \ldots + g_n b_n x_n = 0,$$
$$\ldots\ldots\ldots\ldots\ldots\ldots\ldots\ldots\ldots\ldots\ldots\ldots\ldots$$
$$g_1 l_1 x_1 + g_2 l_2 x_2 + \ldots + g_n l_n x_n = 0,$$

which will be as many as there are unknown quantities $s, t \ldots z$. If in these we place for $x_1, x_2 \ldots x_n$ their values from (74), we have the final normal equations which determine the most probable values of the unknown quantities. As in Art. 33, we abbreviate the expressions of these equations by placing

(79)
$$\Sigma g a^2 = g_1 a_1^2 + g_2 a_2^2 + \ldots + g_n a_n^2,$$
$$\Sigma g a b = g_1 a_1 b_1 + g_2 a_2 b_2 + \ldots + g_n a_n b_n,$$
$$\Sigma g a M = g_1 a_1 M_1 + g_2 a_2 M_2 + \ldots + g_n a_n M_n,$$
$$\text{etc.} \qquad \text{etc.} \qquad \text{etc.}$$

And thus have the normal equations

(80)
$$\Sigma g a^2 . s + \Sigma g a b . t + \ldots + \Sigma g a l . z = \Sigma g a M,$$
$$\Sigma g a b . s + \Sigma g b^2 . t + \ldots + \Sigma g b l . z = \Sigma g b M,$$
$$\ldots\ldots\ldots\ldots\ldots\ldots\ldots\ldots\ldots\ldots\ldots\ldots$$
$$\Sigma g a l . s + \Sigma g b l . t + \ldots + \Sigma g l^2 . z = \Sigma g l M,$$

by whose solution we find the values of $s, t \ldots z$.

If all our observations are of equal weight, we may place $g = 1$, and then our equations (77), (78), (79) and (80) reduce to (68), (70), (72) and (73), as determined for that case by another method.

37. The notation above exhibited is very useful as a guide in dealing with large numbers of observations, and in reducing the numerical operations to a routine for com-

DEVELOPMENT OF PRACTICAL METHODS

puters. To illustrate its uses, suppose we have five observations giving the following equations:

$$2s - 2t + 3u = 5 \text{ with weight } 3,$$
$$s + 2t + u = 9 \quad\ldots\ldots\ldots\ldots\ 2,$$
$$3s - t + u = 12 \quad\ldots\ldots\ldots\ldots\ 5,$$
$$4t - 3u = 5 \quad\ldots\ldots\ldots\ldots\ 1,$$
$$-s - t + 2u = -7 \quad\ldots\ldots\ldots\ldots\ 4,$$

from which we wish to find the most probable values of s, t and u.

Comparing these with (74), we have

$a_1 = 2,$	$b_1 = -2,$	$c_1 = 3,$	$M_1 = 5,$	$g_1 = 3,$
$a_2 = 1,$	$b_2 = 2,$	$c_2 = 1,$	$M_2 = 9,$	$g_2 = 2,$
$a_3 = 3,$	$b_3 = -1,$	$c_3 = 1,$	$M_3 = 12,$	$g_3 = 5,$
$a_4 = 0,$	$b_4 = 4,$	$c_4 = -3,$	$M_4 = 5,$	$g_4 = 1,$
$a_5 = 1,$	$b_5 = -1,$	$c_5 = 2,$	$M_5 = -7,$	$g_5 = 4.$

Then we have

$$a_1^2 = 4, \quad a_1 b_1 = -4, \quad b_1 c_1 = -6, \quad a_4 b_4 = 0, \text{ etc.,}$$

and hence from (79) we form the sums

$$\Sigma g a^2 = g_1 a_1^2 + g_2 a_2^2 + g_3 a_3^2 + g_4 a_4^2 + g_5 a_5^2 = 63,$$
$$\Sigma g a b = g_1 a_1 b_1 + g_2 a_2 b_2 + g_3 a_3 b_3 + g_4 a_4 b_4 + g_5 a_5 b_5 = -27,$$
$$\Sigma g a c = g_1 a_1 c_1 + g_2 a_2 c_2 + g_3 a_3 c_3 + g_4 a_4 c_4 + g_5 a_5 c_5 = 43,$$
$$\Sigma g a M = 200, \quad \Sigma g b^2 = 45, \quad \Sigma g b c = -39,$$
$$\Sigma g b M = -6, \quad \Sigma g c^2 = 59, \quad \Sigma g c M = 52,$$

and inserting these in (80), we have the normal equations

$$63s - 27t + 43u = 200,$$
$$-27s + 45t - 39u = -6,$$
$$43s - 39t + 59u = 52,$$

whose solution will furnish the most probable values of s, t and u. For GAUSS' method of still further employing the notation in the solution of the normal equations, see Art. 60 of the Appendix.

The student will observe that the above process is identical with that of multiplying each observation equation by the square root of its weight, and then forming the normal equations by the method used in Part I.

38. *To determine the probable errors* of the determined quantities $s, t \ldots z$, we let $G_s, G_t \ldots G_z$ denote their weights, and $R_s, R_t \ldots R_z$ their probable errors. Then if r be the probable error of an observation whose *weight is unity*, we have

(53) $$G_s : G_t : 1 :: \frac{1}{R_s^2} : \frac{1}{R_t^2} : \frac{1}{r^2},$$

from which we find

(81) $$R_s = \frac{r}{\sqrt{G_s}}, \quad R_t = \frac{r}{\sqrt{G_t}}, \text{ etc.}$$

Hence in order to find the probable errors of $s, t \ldots z$, we have only to determine the probable error of an observation of the weight 1, and the weights $G_s, G_t \ldots G_z$. And in general, if r_1 be the probable error of a measurement whose weight is g_1, r_1 is equal to r divided by the square root of g_1 (Art. 30).

To find the probable error of an observation whose weight is unity we give the following reasoning.

Suppose that we have formed and solved the normal equations (80), and found the most probable values of the unknown quantities. Let those most probable values be represented by $s, t \ldots z$, and the corresponding true values by $s + \delta s$, $t + \delta t \ldots z + \delta z$, in which $\delta s, \delta t \ldots \delta z$ are small unknown corrections. Now if in (74) we substitute the values $s, t \ldots z$, they will not reduce to zero, but leave the *residuals* $v_1, v_2 \ldots v_n$. Thus

11—2

$$a_1 s + b_1 t + \ldots + l_1 z - M_1 = v_1 \text{ with weight } g_1,$$
(82) $\quad a_2 s + b_2 t + \ldots + l_2 z - M_2 = v_2 \ldots\ldots\ldots\ldots g_2,$
$$\ldots\ldots\ldots\ldots\ldots\ldots\ldots\ldots\ldots\ldots\ldots\ldots\ldots\ldots\ldots\ldots$$
$$a_n s + b_n t + \ldots + l_n z - M_n = v_n \text{ with weight } g_n,$$

while if the corresponding true values are inserted, we shall have the *errors* $x_1, x_2 \ldots x_n$. Thus

$$a_1(s + \delta s) + b_1(t + \delta t) + \ldots - M_1 = x_1,$$
(83) $\quad a_2(s + \delta s) + b_2(t + \delta t) + \ldots - M_2 = x_2,$
$$\ldots\ldots\ldots\ldots\ldots\ldots\ldots\ldots\ldots\ldots\ldots\ldots\ldots\ldots$$
$$a_n(s + \delta s) + b_n(t + \delta t) + \ldots - M_n = x_n.$$

Now each one of the latter equations may be written in the form

$$as + bt + \ldots + lz - M,$$
$$+ a\delta s + b\delta t + \ldots + l\delta z = x,$$

and as (82) are of the form

$$as + bt + \ldots + lz - M = v,$$

the equations (83) will reduce to

$$v_1 + a_1 \delta s + b_1 \delta t + \ldots + l_1 \delta z = x_1 \text{ weight } g_1,$$
(84) $\quad v_2 + a_2 \delta s + b_2 \delta t + \ldots + l_2 \delta z = x_2 \ldots\ldots\ldots g_2,$
$$\ldots\ldots\ldots\ldots\ldots\ldots\ldots\ldots\ldots\ldots\ldots\ldots\ldots\ldots$$
$$v_n + a_n \delta s + b_n \delta t + \ldots + l_n \delta z = x_n \text{ weight } g_n.$$

Now the principle of least squares requires that the expression

(76) $\quad g_1 x_1^2 + g_2 x_2^2 + \ldots + g_n x_n^2 = \Sigma g x^2$

shall be made a minimum to give the most probable values of $s, t \ldots z$, and by the solution of the normal equations we have for its minimum value the sum $\Sigma g v^2$. From the residual equations (84), we may find a relation connecting the two sums $\Sigma g v^2$ and $\Sigma g x^2$ by squaring both members of each of those equations, multiplying each by its corresponding weight, and then adding the results. Without actually performing these operations, we see that if the squares and products of $\delta s, \delta t \ldots \delta z$ be neglected as small in comparison with $\delta s, \delta t \ldots \delta z$, the result will be of the form

$$(85) \qquad \Sigma g v^2 + k_1 \delta s + k_2 \delta t + \ldots + k_q \delta t = \Sigma g x^2,$$

in which $k_1, k_2 \ldots k_q$ are coefficients of the unknown corrections, and dependent only upon the known constants a_1, b_1, etc. If the number of unknown quantities is q there will be q of these terms. Placing

$$k_1 \delta s = k_s^2, \qquad k_2 \delta t = k_t^2, \text{ etc.,}$$

we write the relation

$$(85) \qquad \Sigma g v^2 + k_s^2 + k_t^2 + \ldots + k_z^2 = \Sigma g x^2.$$

Now the probability of the occurrence of the error x_1 whose measure of precision is h_1, and whose weight is g_1, is by (19) and (57),

$$(58) \qquad y_1 = h \sqrt{g_1} \, i_1 \pi^{-\frac{1}{2}} e^{-h^2 g_1 x_1^2},$$

in which h is the measure of precision of an observation of the weight 1. And hence by exactly the same reasoning as in Art. 30 we may show that when n is a large number,

$$(61) \qquad \Sigma g x^2 = \frac{n}{2h^2}.$$

Further, if we suppose that all the q unknown quantities $s, t \ldots z$ except *one* are zero, the equation (85) being true for any number q, will hold good for $q = 1$, and will contain then only one k^2, whose value as shown in Art. 30 is $\frac{1}{2h^2}$. Hence since each of these k^2 is a constant, the value of each must be $\frac{1}{2h^2}$ whatever be their number. Hence the equation (85) is

$$\Sigma g v^2 + \frac{1}{2h^2} + \frac{1}{2h^2} + \ldots + \frac{1}{2h^2} = \frac{n}{2h^2},$$

and since there are q terms whose value is $\frac{1}{2h^2}$, it is

(85) $$\Sigma g v^2 + \frac{q}{2h^2} = \frac{n}{2h^2},$$

from which we find

(86) $$h = \sqrt{\frac{n-q}{2\Sigma g v^2}}.$$

Therefore from the constant relation between h and r (28) the probable error of an observation of the weight unity is

(87) $$r = 0\cdot 6745 \sqrt{\frac{\Sigma g v^2}{n-q}}.$$

In the above we have followed closely the reasoning of Art. 30. The student can also readily apply the method of Art. 24 to produce the same result, by finding from (58) the product P of the simultaneous existence of the errors $x_1, x_2 \ldots x_n$, and then determining h by differentiation.

If in (87) we make $q = 1$, or consider only one measured quantity, it reduces to formula (65), and if we make $g = 1$, or regard the observations as of equal weight, it becomes (42). Thus measurements upon one quantity are but a special case of the more general one of indirect observations.

39. The probable errors of the values of $s, t \ldots z$ can now be found from (81) as soon as the weights G_s, G_t, etc., are known. We now proceed to determine these.

The observations (74) upon the quantities $M_1, M_2 \ldots M_n$ furnish the normal equations (80). The solution of these equations gives the values of $s, t \ldots z$ in terms of $M_1, M_2 \ldots M_n$ and coefficients independent of those quantities. Suppose the general solution to give

(88)
$$s = \sigma_1 M_1 + \sigma_2 M_2 + \sigma_3 M_3 + \ldots + \sigma_n M_n,$$
$$t = \tau_1 M_1 + \tau_2 M_2 + \tau_3 M_3 + \ldots + \tau_n M_n,$$
$$\ldots\ldots\ldots\ldots\ldots\ldots\ldots\ldots\ldots\ldots\ldots\ldots\ldots\ldots$$
$$z = \zeta_1 M_1 + \zeta_2 M_2 + \zeta_3 M_3 + \ldots + \zeta_n M_n,$$

in which the coefficients $\sigma, \tau \ldots \zeta$ depend only upon the constants $a, b \ldots l$ and the weights g in (80). Then if R_s is the probable error of s and $r_1, r_2 \ldots r_n$ the probable errors of $M_1, M_2 \ldots M_n$, we have, by a principle to be proved in Art. 42, since the measurements are independent,

$$(102) \quad R_s^2 = \sigma_1^2 r_1^2 + \sigma_2^2 r_2^2 + \ldots + \sigma_n^2 r_n^2 = \Sigma \sigma^2 r^2.$$

Now G_s being the weight of s, and $g_1, g_2 \ldots g_n$ the weights of the measurements $M_1, M_2 \ldots M_n$, we have from (53) and (54)

$$(89) \quad r_1^2 = \frac{G_s R_s^2}{g_1}, \quad r_2^2 = \frac{G_s R_s^2}{g_2} \ldots r_n^2 = \frac{G_n R_n^2}{g_n},$$

and by substitution in equation

$$(90) \quad R_s^2 = G_s R_s^2 \left(\frac{\sigma_1^2}{g_1} + \frac{\sigma_2^2}{g_2} + \ldots + \frac{\sigma_n^2}{g_n} \right) = G_s R_s^2 \Sigma \frac{\sigma^2}{g},$$

from which we find

$$(91) \quad G_s = \frac{1}{\Sigma \dfrac{\sigma^2}{g}},$$

in which the usual notation for sums is followed. In like manner we may show that the weight of t_1 is the reciprocal of $\Sigma \dfrac{\tau^2}{g}$, and that the weight of z is the reciprocal of $\Sigma \dfrac{\zeta^2}{g}$.

Owing however to the labour of finding the coefficients $\sigma, \tau \ldots \zeta$ it is better to deduce these expressions under a different form. Let us suppose the normal equations (80) to be solved, giving

$$(92) \quad \begin{aligned} s &= \alpha_1 \Sigma gaM + \alpha_2 \Sigma gbM + \ldots + \alpha_q \Sigma glM, \\ t &= \beta_1 \Sigma gaM + \beta_2 \Sigma gbM + \ldots + \beta_q \Sigma glM, \\ & \ldots\ldots\ldots\ldots\ldots\ldots\ldots\ldots\ldots\ldots\ldots\ldots\ldots\ldots \\ z &= \lambda_1 \Sigma gaM + \lambda_2 \Sigma gbM + \ldots + \lambda_q \Sigma glM, \end{aligned}$$

in which $\alpha, \beta \ldots \lambda$ are coefficients independent of $M_1, M_2 \ldots M_n$. Then the respective weights of $s, t \ldots z$ will be $\frac{1}{\alpha_1}, \frac{1}{\beta_2} \ldots \frac{1}{\lambda_q}$ (z being the last unknown quantity and q the number of unknown quantities). In order to prove this let us find the weight of t. By comparison of (88) and (92) we have

(93)
$$\tau_1 = \beta_1 g_1 a_1 + \beta_2 g_1 b_1 + \ldots + \beta_q g_1 l_1,$$
$$\tau_2 = \beta_1 g_2 a_2 + \beta_2 g_2 b_2 + \ldots + \beta_q g_2 l_2,$$
$$\ldots\ldots\ldots\ldots\ldots\ldots\ldots\ldots\ldots\ldots\ldots\ldots$$
$$\tau_n = \beta_1 g_n a_n + \beta_2 g_n b_n + \ldots + \beta_q g_n l_n.$$

Squaring each of these equations, dividing each by its g and adding the results, we have

(94)
$$\Sigma \frac{\tau^2}{g} = \beta_1 (\beta_1 \Sigma g a^2 + \beta_2 \Sigma g a b + \ldots + \beta_q \Sigma g a l)$$
$$+ \beta_2 (\beta_1 \Sigma g a b + \beta_2 \Sigma g b^2 + \ldots + \beta_q \Sigma g b l) + \ldots$$
$$+ \beta_q (\beta_1 \Sigma g a l + \beta_2 \Sigma g b l + \ldots + \beta_q \Sigma g l^2).$$

Now if we were to solve the normal equations (80) by the method of indeterminate multipliers (Art. 35), we might multiply the first by a number β_1, the second by β_2, the qth by β_q, and add the results; then if upon these multipliers we impose the conditions

(95)
$$\beta_1 \Sigma g a^2 + \beta_2 \Sigma g a b + \ldots + \beta_q \Sigma g a l = 0,$$
$$\beta_1 \Sigma g a b + \beta_2 \Sigma g b^2 + \ldots + \beta_q \Sigma g b l = 1,$$
$$\ldots\ldots\ldots\ldots\ldots\ldots\ldots\ldots\ldots\ldots\ldots\ldots$$
$$\beta_1 \Sigma g a l + \beta_2 \Sigma g b l + \ldots + \beta_q \Sigma g l^2 = 0,$$

all the terms except those involving t will reduce to zero, and the value of t will be the same as given by the second of equations (92). Comparing then (94) and (95), we have

(96) $\quad \Sigma \dfrac{\tau^2}{g} = \beta_2,\quad$ or $\quad \dfrac{1}{\Sigma \dfrac{\tau^2}{g}} = \dfrac{1}{\beta_2} = G_t.$

Hence the weight of t is $\dfrac{1}{\beta_2}$, which was to be proved.

Although we have here supposed the solution to be made by the method of indeterminate multiplers, it is evident that the same result will be reached whatever be the method employed. Therefore to find the weights of the values of the unknown quantities, we have only to solve the normal equations preserving the absolute terms in literal form; then the weight of s is the reciprocal of the coefficient of the absolute term in the normal equation for s occurring in the general value of s, the weight of t is the reciprocal of the corresponding absolute term in the general value of t, and so on.

40. Thus in the example of Art. 35 we have found the general values of y and s. The weight of y is then $\frac{17}{7}$, the reciprocal of the coefficient of E in the value of y; and the weight of s is $\frac{51}{32}$, the reciprocal of the coefficient of A in the value of s.

If there be but one unknown quantity, the observation equations will be

$$a_1 z = M_1, \quad a_2 z = M_2 \ldots a_n z = M_n,$$

and if the weights of these be $g_1, g_2 \ldots g_n$, the single normal equation is $\Sigma g a^2 . z = \Sigma g a M$. If $a_1 = a_2 = 1$, the observations are made directly upon M, and the equation gives $z = \frac{\Sigma g M}{\Sigma g}$, which agrees with the general mean (50). By Art. 39 the weight of z must be the reciprocal of $\frac{1}{\Sigma g}$, or

$$\Sigma g = g_1 + g_2 + \ldots + g_n,$$

as shown otherwise in Art. 27. Further, if $g = 1$ the normal equation becomes $nz = \Sigma M_1$, agreeing with the law of the arithmetical mean (27). Thus is the correctness of our methods verified by these mutual checks.

---

41. *The probable errors of functions of independently observed quantities* will next be investigated. Let us take

first the most simple function of two independently measured quantities, viz.
$$Z = z_1 \pm z_2,$$
in which z_1 and z_2 are the measured quantities; then it is evident that if z_1 and z_2 are the most probable values found, the most probable value of Z is the sum or difference of z_1 and z_2, as the case may be. Let the errors arising in the measurements be

for z_1; x_1', x_1'', x_1''', etc.;

for z_2; x_2', x_2'', x_2''', etc.;

then the errors of Z will be

$$X_1 = x_1' \pm x_2', \qquad X_2 = x_1'' \pm x_2'', \qquad X_3 = x_1''' \pm x_2''', \text{ etc.}$$

Squaring and adding these errors, we have

$$X_1^2 + X_2^2 + \text{etc.} = (x_1' \pm x_2')^2 + (x_1'' \pm x_2'')^2 + \text{etc.},$$

or developing and adding

(97) $$\Sigma X^2 = \Sigma x_1^2 + 2\Sigma x_1 x_2 + \Sigma x_2^2.$$

In a great number of observations there will probably be as many positive as negative products of the form x_1, x_2, and hence we shall have $\Sigma x_1 x_2 = 0$. Hence (97) becomes

(98) $$\Sigma X^2 = \Sigma x_1^2 + \Sigma x_2^2.$$

Denoting the measures of precision of Z, z_1 and z_2 by H, h_1 and h_2, this becomes by (34)

(98) $$\frac{n}{2H^2} = \frac{n}{2h_1^2} + \frac{n}{2h_2^2}.$$

Denoting the probable errors by R, r_1 and r_2, we have

(28) $$HR = h_1 r_1 = h_2 r_2 = 0.4769,$$

and hence this relation in connection with (98) gives us

(99) $$R^2 = r_1^2 + r_2^2.$$

In like manner, if we have a function of the sum or difference of several independent observed quantities, viz.

$$Z = z_1 \pm z_2 \pm \ldots \pm z_n,$$

the probable error of Z will be given by the relation

(100) $\qquad R^2 = r_1^2 + r_2^2 + r_3^2 + \ldots + r_n^2.$

42. Next let Z be a function connected with an observed quantity z_1, by the constant relation

$$Z = A z_1.$$

Then if the probable error of z_1 is r_1, it is evident, since an error x_1 in the measurement of z_1 gives to Z an error $A x_1$, that the probable error of Z is

(101) $\qquad R = A r_1, \text{ or } R^2 = A^2 r_1^2.$

Hence combining this with the principle deduced in the previous Article, if

$$Z = A z_1 + B z_2 + C z_3 + \text{etc.}$$

and if z_1, z_2, etc. are *independently* observed with the probable errors r_1, r_2, r_3, etc., the probable error of Z is given by

(102) $\qquad R^2 = A^2 r_1^2 + B^2 r_2^2 + C^2 r_3^2 + \text{etc.}$

Thus if $z_1, z_2, z_3 \ldots z_n$ are several observed values of the *same* quantity, the probable error of their sum is

$$R = \sqrt{r_1^2 + r_2^2 + \ldots + r_n^2} = \sqrt{n r^2},$$

and by (101) the probable error of $\frac{1}{n}$th of this sum is

$$r_0 = \frac{\sqrt{n r^2}}{n} = \frac{r}{\sqrt{n}},$$

which is the probable error of the arithmetical mean, as has been otherwise shown in Art. 26.

Conditioned Observations.

43. In all that precedes we have supposed that the quantities to be determined by observation were independent of each other. Although they have been related to each other through the observation equations, and have been required to satisfy approximately those equations, they have been so far independent, that any one unknown quantity might be supposed to vary without affecting the values of the others. The methods above developed show how, out of the many equal *possible* systems of values, we can determine the most probable.

We now come to a second class of observations in which all systems of values are not equally possible owing to the existence of conditions which must be exactly satisfied. Thus having measured two angles of a triangle, the adjusted value of one is entirely independent of that of the other, but if the third angle be measured, the three angles are subject to the rigorous geometrical condition that their sum must be exactly $180°$. We have then in conditioned observations two classes of equations, observation equations and *conditional equations*, the number of the first being generally greater than the number of unknown quantities, and that of the latter always less.

44. The number of observation equations we designate as before by n, the number of unknown quantities by q, and the number of conditional equations by p. If no conditional equations existed, the principle of least squares (Art. 14) would require that the adjusted system of values should be the most probable for the n independent observation equations. But here these n equations are conditioned by p conditional equations. The p conditional equations being less in number than the q unknown quantities, may be satisfied in various ways, and further, the final adjusted system of values must exactly satisfy them. Hence we conclude that *of all the systems of values which exactly satisfy the p conditional equations, that one is to be chosen as the best* *which in the n observation equations makes the sum of the squares of the residuals a minimum.*

We may then reduce the problem of conditioned observations to that of independent ones, by finding from the ρ conditional equations the values of q unknown quantities in terms of the remaining $q-\rho$ quantities and substituting them in the n observation equations. There will thus result n observation equations, containing, however, only $q-\rho$ instead of q unknown quantities, and each of these equations will represent an *independent* observation. From these equations we proceed to form the normal equations (73), whose solution will give us the most probable values of the $q-\rho$ unknown quantities. Substituting these values in the ρ conditional equations, we find the values of the remaining q unknown quantities. Thus the system of values will exactly satisfy the conditional equations, and at the same time be the most probable system for the observation equations. This, therefore, is a general solution of the problem.

45. Although this is perfectly general and simple in theory, it gives rise in practice to tedious computations, for we have first from the ρ conditional equations to eliminate q unknown quantities, and again solve the normal equations to find the values of the remaining $q-\rho$ quantities. The process generally used by computers is GAUSS' *Method of Correlatives*, which in Art. 45 of Part I. is fully illustrated by examples of conditioned observations of equal weight, and which in Art. 48 we shall proceed to develope for the more general case of unequal weights. To apply the proof to those of equal precision or weight, we have only to place $g = 1$, that is, omit g from the formulæ.

47. The method of Art. 44 is perfectly general, whether the observations be of equal or unequal weight. We have in the latter case to find from the ρ conditional equations the values of any q unknown quantities in terms of the remaining $q-\rho$ quantities, and substitute their values in the n observation equations (74), each of which is then independent; then, applying the weights we form the normal equations (80), whose solution gives us the best system of values for the $q-\rho$ quantities. The remaining ρ quantities

are then directly found from the ρ conditional equations. The method of correlatives is however greatly to be preferred for simplicity in the numerical operations; and this we now proceed to develope for the ordinary case of linear conditional equations, to which all others may always be reduced.

48.* Let n observations be made to determine the values of q unknown quantities, which are subject to ρ rigorous conditions. Whether the measurements be direct or indirect, let them first be supposed independent, and let them be adjusted by the methods of Chapters II. and III., and let the resulting values be $S, T \ldots Z$, having the weights $g_0, g_1 \ldots g_x$. Let the most probable system of values which we are to find be $s, t \ldots z$, and the conditional equations which they are to exactly satisfy be

(103)
$$\alpha_1 s + \alpha_2 t + \ldots + \alpha_q z = N_1,$$
$$\beta_1 s + \beta_2 t + \ldots + \beta_q z = N_2,$$
$$\ldots\ldots\ldots\ldots\ldots\ldots\ldots\ldots\ldots\ldots\ldots$$
$$\lambda_1 s + \lambda_2 t + \ldots + \lambda_q z = N_\rho.$$

Then if the values $S, T \ldots Z$ exactly satisfy these equations no further adjustment is necessary. If not, let $s', t' \ldots z'$ be a system of corrections which applied to $S, T \ldots Z$ will make them equal to $s, t \ldots z$, so that

$$s = S + s', \quad t = T + t' \ldots z = Z + z'.$$

Then substituting these in (103) they reduce to

(104)
$$\alpha_1 s' + \alpha_2 t' + \ldots + \alpha_q z' = N_1 - (\alpha_1 S + \alpha_2 T + \ldots) = N',$$
$$\beta_1 s' + \beta_2 t' + \ldots + \beta_q z' = N_2 - (\beta_1 S + \beta_2 T + \ldots) = N'',$$
$$\ldots\ldots\ldots\ldots\ldots\ldots\ldots\ldots\ldots\ldots\ldots\ldots\ldots\ldots\ldots$$
$$\lambda_1 s' + \lambda_2 t' + \ldots + \lambda_q z' = N_\rho - (\lambda_1 S + \lambda_1 T + \ldots) = N^{(\rho')},$$

* Essentially the demonstration given by CHAUVENET in his excellent *Treatise on Least Squares*, the Appendix to his *Spherical and Practical Astronomy*, Philadelphia, 1867.

in which $s', t' \ldots z'$ are the unknown quantities, and N', N'', etc. constants depending upon the theoretical constants N_1, N_2, etc. and the observed values $S, T \ldots Z$. The number of these equations is ρ.

Now the n approximate observation equations are
$$s = S, \quad t = T \ldots z = Z,$$
which we may write
(105) $\quad s - S = s' = 0, \; t - T = t' = 0 \ldots z - Z = z' = 0,$

whose respective weights are $g_s, g_t \ldots g_z$; and the values which we are to determine for $s', t' \ldots z'$ must not only exactly satisfy the conditions (104), but also be the most probable set of values for (105). Since $s', t' \ldots z'$ are themselves the residuals in the observation equations, this latter requirement is by our fundamental principle (26) satisfied when the quantity

(76) $\quad \Sigma g v^2 = g_s s'^2 + g_t t'^2 + \ldots + g_z z'^2 =$ a minimum.

Putting then the differential of this quantity equal to zero, we have

(106) $\quad g_s s' ds' + g_t t' dt' + \ldots + g_z z' dz' = 0.$

Now if $s', t' \ldots z'$ were *independent* of each other, the differential coefficient of $\Sigma g v^2$ (76) with respect to each of the variables would necessarily be zero (as in Art. 36); and then $s', t' \ldots z'$ being each zero, the most probable values of $s, t \ldots z$ would be their adjusted or observed values $S, T \ldots Z$. But this expression (106) is conditioned by the equations (104), and no values of $s, t \ldots z$ can be admitted which do not exactly satisfy those equations. If then we differentiate (104) we have the equations

(107) $\quad \begin{aligned} & a_1 ds' + a_2 dt' + \ldots + a_q dz' = 0, \\ & \beta_1 ds' + \beta_2 dt' + \ldots + \beta_q dz' = 0, \\ & \ldots\ldots\ldots\ldots\ldots\ldots\ldots\ldots\ldots\ldots\ldots\ldots \\ & \lambda_1 ds' + \lambda_2 dt' + \ldots + \lambda_q dz' = 0, \end{aligned}$

with which (106) must agree and coexist.

The number of the equations (107) is ρ, the number of the differentials ds', dt' ... dz' is q, and since from the nature of the case q is greater than ρ, we can from (107) find the values of ρ differentials in terms of the remaining $q-\rho$ differentials. Let us suppose this elimination to be performed, and that the values of ρ differentials found in terms of the others are then substituted in (106); we shall thus have an equation in which the remaining $q-\rho$ differentials will be independent, and the coefficients of each of these $q-\rho$ differentials will therefore be severally equal to zero. Without actually performing this process in each particular case we can arrive at the general result of such an elimination and substitution as follows. Multiply the first equation of (107) by the indeterminate number K_1, the second by K_2, ... the last by K_ρ, and also the equation (106) by -1, and form the sum of these products. Then if K_1, K_2 ... K_ρ be determined so that ρ differentials shall disappear (Art. 35), the final equation will contain only the remaining $q-\rho$ differentials. But these being independent, their coefficients will be severally equal to zero; and hence we have in all q conditions that the coefficients ds', dt' ... dz' in that sum shall be each equal to zero, viz.

(108)
$$\alpha_1 K_1 + \beta_1 K_2 + \ldots + \lambda_1 K_\rho - g_s s' = 0,$$
$$\alpha_2 K_1 + \beta_2 K_2 + \ldots + \lambda_2 K_\rho - g_t t' = 0,$$
$$\ldots\ldots\ldots\ldots\ldots\ldots\ldots\ldots\ldots\ldots\ldots\ldots\ldots\ldots$$
$$\alpha_q K_1 + \beta_q K_2 + \ldots + \lambda_q K_\rho - g_z z' = 0.$$

If now we multiply the first of these by $\dfrac{\alpha_1}{g_s}$, the second by $\dfrac{\alpha_2}{g_t}$, ... the last by $\dfrac{\alpha_q}{g_z}$, and add the products, we have by comparison with the first equation of (104),

$$\Sigma \frac{\alpha^2}{g} \cdot K_1 + \Sigma \frac{\alpha\beta}{g} \cdot K_2 + \text{etc.} = N',$$

in which we have the usual notation for sums, or

$$\Sigma \frac{\alpha^2}{g} = \frac{\alpha_1^2}{g_s} + \frac{\alpha_2^2}{g_t} + \ldots + \frac{\alpha_q^2}{g_s}$$

$$\Sigma \frac{\alpha\beta}{g} = \frac{\alpha_1\beta_1}{g_s} + \frac{\alpha_2\beta_2}{g_t} + \ldots + \frac{\alpha_q\beta_q}{g_s}, \text{ etc.}$$

In like manner multiplying the first by $\frac{\beta_1}{g_s}$, the second by $\frac{\beta_2}{g_t}$, etc. we form a second normal equation. Thus we have in all p normal equations containing the p new auxiliary unknown quantities, viz.

$$\Sigma \frac{\alpha_1^2}{g} \cdot K_1 + \Sigma \frac{\alpha\beta}{g} \cdot K_2 + \ldots + \Sigma \frac{\alpha\lambda}{g} K_p = N'$$

(109) $$\Sigma \frac{\alpha\beta}{g} \cdot K_1 + \Sigma \frac{\beta^2}{g} \cdot K_2 + \ldots + \Sigma \frac{\beta\lambda}{g} K_p = N''$$

. . . .

$$\Sigma \frac{\alpha\lambda}{g} \cdot K_1 + \Sigma \frac{\beta\lambda}{g} \cdot K_2 + \ldots + \Sigma \frac{\lambda^2}{g} \cdot K_p = N^{(\rho)}.$$

The solution of these equations will give the values of $K_1, K_2 \ldots K_p$, which being substituted in the *correlative equations* (108) will furnish the values of the required corrections $s', t' \ldots z'$; thus

$$s' = \frac{1}{g_s}(\alpha_1 K_1 + \beta_1 K_2 + \ldots + \lambda_1 K_p),$$

(110) $$t' = \frac{1}{g_t}(\alpha_2 K_1 + \beta_2 K_2 + \ldots + \lambda_2 K_p),$$

etc. etc.

and these values will not only exactly satisfy the conditional equations (104) but will make the sum $\Sigma g v^2$ a minimum, and hence are the best attainable system of values. Adding these corrections to the observed values $S, T \ldots Z$, we have finally the adjusted values $s, t \ldots z$, and these will exactly satisfy the conditional equations (103).

M.

49. *The Probable Errors* of conditioned observations follow directly from the formulæ of Art. 38, and the general solution of Art. 44.

Since the n observation equations contain q unknown quantities, and by elimination from the p conditional equations we reduce that number to $q - p$ independent unknown quantities, we have only in the formula (87) to replace q by $q - p$. Therefore the probable error of an observation *of the weight unity* is

$$(111) \qquad r = 0.6745 \sqrt{\frac{\Sigma g v^2}{n - q + p}},$$

and the probable errors of observations or values whose weights are g_1, G_s, etc. are

$$(81) \qquad r_1 = \frac{r}{\sqrt{g_1}}, \quad R_s = \frac{r}{\sqrt{G_s}}, \text{ etc.}$$

If we have determined the probable values of $s, t \ldots z$ by means of corrections $s', t' \ldots z'$ to the observed values $S, T \ldots Z$, the sum $\Sigma g v^2$ is simply

$$\Sigma g v^2 = g_s s'^2 + g_t t'^2 + \ldots + g_z z'^2,$$

in which $g_s, g_t \ldots g_z$ are given by the observations themselves or by adjustment as in Arts. 28 and 39. The final weights G_s, G_t, etc. of the values s, t, etc. are found by exactly the same process as developed in Art. 39. An example illustrating the operation in full is given in Part I.

The Discussion of Observations.

50. The above methods constitute the whole science of Least Squares as applied to observations involving linear equations, to which all others may be reduced (Art. 59); and the most common formulæ and methods for investigating probable errors. The determination of probable errors of functions of dependent quantities has not been given, as such investigations are rarely needed in practical discussions, and would be out of place in an elementary text-book. A few points in Chapter V. require perhaps further notice and explanation.

51. *The Deduction of Empirical Formulæ* by the discussion of observations, is one of the most extensive applications of the foregoing methods. Having given the numerical results of a series of physical observations, we have only to assume a general algebraic expression, which includes the law of the phenomena in terms of the observed quantities and undetermined constants. Then inserting the numerical results, we have a series of observation equations from which we deduce the normal equations, whose solution gives the most probable values of the constants. The choosing of the proper algebraic expression is usually the most difficult part of this process. For this, no general rules can be given. The best plan is to assume convenient horizontal and vertical units and plot the results of the observations, thus obtaining a curve which represents them graphically to the eye. A comparison of this curve with similar curves whose equations are known, will then often enable us to determine the general form of a convenient algebraic expression.

52. If the plotted curve resembles a parabola or hyperbola, it may be represented by the equation

$$(112) \qquad y = A + Bx + Cx^2 + \text{etc.}$$

in which the absolute term A may often be directly determined by choosing a proper origin for the values of y and x.

53. If however the plotted curve repeats itself like the curve of signs, the general equation

$$(113) \quad y = A + B \sin\left(\frac{360°}{m} x + B'\right) + C \sin\left(\frac{360}{m} 2x + C'\right) + \text{etc.}$$

will be applicable. Here also the constants A, B', C' may often be omitted by choosing a proper point as the origin of the co-ordinates x and y. The value of m is generally to be assumed from the inspection of the plotted curves, or its probable value be found by successive approximations. If this formula be expanded, we have, considering only the terms involving A and B,

$$y = A + B \sin \frac{360°}{m} x \cos B' + B \cos \frac{360°}{m} x \sin B' + \text{etc.},$$

and if in this we place

$$B \cos B' = B_1, \qquad B \sin B' = B_2,$$

it becomes

(114) $\qquad y = A + B_1 \sin \frac{360°}{m} x + B_2 \cos \frac{360°}{m} x + \text{etc.}$

which is a more convenient form for computation. Inserting in this the values of y and x from the observations, we form the normal equations, and deduce the probable values of A, B_1, and B_2. Then by means of the above relations which furnish

$$B = \sqrt{B_1^2 + B_2^2}, \quad \cos B' = \frac{B_1}{B}, \text{ and } \sin B' = \frac{B_2}{B},$$

the derived equation (114) can, if desired, be reduced to the form (113), which is often more convenient for subsequent discussion.

54. Formula involving undetermined constants like the case of the pendulum given in Part I. occasionally arise in theoretical investigations; and if observations enough exist, the constants may be deduced.

The determination of the probable errors of such formulæ is rarely necessary, as the comparison of the computed and observed results indicate their precision sufficiently well to enable us to decide upon the degree of confidence to which they are entitled. The weights and probable errors of the deduced constants can in all cases be found by the methods of Arts. 38 and 39. The probable errors of the results deduced from such formula (for example, the probable errors of the values found for y in (113) after A, B, etc. have been determined) cannot however be found from the relation given in (102) because the separate terms are *not independent*. For methods applicable to such cases we must refer the reader to the larger and more complete treatises upon the subject, a list of which is given in Art. 64.

55. *The discussion of the probability of errors* or of the accuracy of observations is of importance in delicate measurements. This we have hitherto done by means of the Probable Error, or the error such that is an even wager that the result is within that amount of the truth (Art. 16). It is perhaps unfortunate that this particular error has been chosen as the one for comparison, for the mind is better satisfied with considering an error such that the probability of an error being less than it is $\frac{99}{100}$ or some higher fraction instead of $\frac{1}{2}$. Such comparisons are readily made by the Table given in Part I., and we have only to explain the manner in which it is calculated. In Art. 12 we have shewn that the expression

$$(22) \qquad P' = \frac{2}{\sqrt{\pi}} \int_0^{hx} e^{-h^2x^2} d.hx,$$

expresses the probability that an error will be included between the limits $-x$ and $+x$. If in this we place

$$hx = \frac{0\cdot 4769 x}{r} = t,$$

we may write it

$$(115) \qquad P' = \frac{2}{\sqrt{\pi}} \int_0^t e^{-t^2} dt,$$

and by the methods explained in the foot-note to equation (22), its value may be found for successive numerical values of $t = hx$. But if we wish instead of h to employ r, we can compute it for successive values of $\frac{x}{r} = \frac{hx}{0\cdot 4769}$, that is, in the Table in Art. 13, Part I, we have only to divide the numbers in the column hx by $0\cdot 4769$ in order to reduce them to the values $\frac{x}{r}$. Then by interpolation the Table is easily written as in Art. 55.

If then we have made n observations which give a mean z_0 with a probable error r_0, we can easily find the probability that z_0 is within $\pm x$ of the truth by taking from the table the fraction corresponding to $\frac{x}{r_0}$. And conversely, if we ask what is the error x, such that it is a wager of 99 to 1 that z_0 is comprised within the limits $z - x$ and $z + x$, we have only to take the number $\frac{x}{r_0}$ corresponding to $P' = 0\cdot 99$.

56. The table may also be used to investigate the probability of constant errors, and to discuss numerous questions arising in the study of statistics, into which, however, the plan of our book forbids us to enter.

APPENDIX.

58. The elementary applications and the theory of Least Squares has now been given and exemplified. A few other applications and extensions valuable to the computer, and a brief notice of the history and literature of the subject, interesting to all who have studied the science, will next be presented.

Observations involving non-linear Equations.

59. In all that precedes, we have supposed that the observations can be represented by equations of the first degree: if this is not the case, but higher equations are involved, they can readily be reduced to linear ones by the following method.

Let the quantities to be determined be represented by $s, t \ldots z$, and the measured quantities by $M_1, M_2 \ldots M_n$, and the observation equations have the general forms,

$$f_1(s, t \ldots z) = M_1,$$
$$f_2(s, t \ldots z) = M_2,$$
$$\ldots\ldots\ldots\ldots\ldots\ldots$$
$$f_n(s, t \ldots z) = M_n,$$

n being the number of observations. These may be written

(116)
$$\phi_1 = f_1(s, t \ldots z) - M_1 = 0,$$
$$\phi_2 = f_2(s, t \ldots z) - M_2 = 0,$$
$$\ldots\ldots\ldots\ldots\ldots\ldots$$
$$\phi_n = f_n(s, t \ldots z) - M_n = 0.$$

APPENDIX.

Now let approximate values of $s, t \ldots z$ be found either by trial or by a solution of a sufficient number of these equations, and let them be denoted by $S, T \ldots Z$, and let $s', t' \ldots z'$ be the most probable system of corrections to these values, so that

$$s = S + s', \quad t = T + t' \ldots z = Z + z'.$$

Developing then the expressions (116) by TAYLOR's theorem, we have, neglecting the products and higher powers of the corrections $s', t' \ldots z'$,

$$\phi_1 = f_1(S, T \ldots Z) - M_1 + \frac{d\phi_1}{ds}s' + \frac{d\phi_1}{dt}t' + \ldots + \frac{d\phi_1}{dz}z',$$

$$\phi_2 = f_2(S, T \ldots Z) - M_2 + \frac{d\phi_2}{ds}s' + \frac{d\phi_2}{dt}t' + \ldots + \frac{d\phi_2}{dz}z',$$

(117) ....

$$\phi_n = f_n(S, T \ldots Z) - M_n + \frac{d\phi_n}{ds}s' + \frac{d\phi_n}{dt}t' + \ldots + \frac{d\phi_n}{dz}z'.$$

Designating the constant term $f_1(S, T \ldots Z)$ by N_1 etc., these become

$$\frac{d\phi_1}{ds}s' + \frac{d\phi_1}{dt}t' + \ldots + \frac{d\phi_1}{dz}z' = M_1 - N_1,$$

(118) $\quad \dfrac{d\phi_2}{ds}s' + \dfrac{d\phi_2}{dt}t' + \ldots + \dfrac{d\phi_2}{dz}z' = M_1 - N_1,$

etc. etc.

where $\dfrac{d\phi_1}{ds}, \dfrac{d\phi_1}{dt}$, etc. are simply the differential coefficients found by differentiating each of the equations (116) with reference to each of the variables and then substituting $S, T \ldots Z$ for $s, t \ldots z$, and are hence constants. Denoting them then by a_1, b_1, etc., we have

$$a_1 s' + b_1 t' + \ldots + l_1 z' = M_1 - N_1,$$

(119) $\quad a_2 s' + b_2^2 t' + \ldots + l_2 z' = M_2 - N_1,$

....

$$a_n s' + b_n t' + \ldots + l_n z' = M_n - N_n.$$

in which all the letters except $s', t' \ldots z'$ denote known quantities. These equations are exactly like those of (66) or (74), and from them we form the normal equations, whose solution gives us the most probable values of the corrections $s', t' \ldots z'$, and hence the best system of values for the observed quantities $s, t \ldots z$.

If non-linear conditional equations are also given, we have only to find approximate values for the unknown quantities, and assume a system of corrections. Then the functional conditional equations may be developed as above by TAYLOR's theorem, and reduced to linear equations of the same form as (104), which may be treated by the method of correlatives and the most probable system of corrections determined, which applied to the approximate values will give the adjusted results. If these do not satisfy the original conditional equations with sufficient accuracy, a new system of corrections may be assumed and the process again repeated.

In Art. 46 is exhibited the reduction of a transcendental conditional equation to a linear one, by the use of the tabular logarithmic differences, which is more convenient than the treatment by TAYLOR's theorem. The latter, however, must be used for higher algebraic or exponential equations.

Gauss' Method for the Solution of Normal Equations.

60. The formation and solution of normal equations is the most laborious part of the practical reduction of observations. In dealing with large numbers of these equations, computers usually follow the method of GAUSS, by which the work is reduced to a systematic routine. This method consists in solving the equations by substitution so as to preserve throughout the work the symmetry which exists in the coefficients of the normal equations. To illustrate it, it will be sufficient to consider a case involving but three unknown quantities arising from observations of equal weight. Let the n observation equations be

(66)
$$a_1 s + b_1 t + c_1 u = M_1,$$
$$a_2 s + b_2 t + c_2 u = M_2, \text{ etc.}$$

APPENDIX.

The three normal equations formed from these will be

(73)
$$\Sigma a^2 . s + \Sigma ab . t + \Sigma ac . u = \Sigma aM,$$
$$\Sigma ab . s + \Sigma b^2 . t + \Sigma bc . u = \Sigma bM,$$
$$\Sigma ac . s + \Sigma bc . t + \Sigma c^2 . u = \Sigma cM,$$

in which

(72)
$$\Sigma b^2 = b_1^2 + b_2^2 + \ldots + b_n^2,$$
$$\Sigma bc = b_1 c_1 + b_2 c_2 + \ldots + b_n c_n, \text{ etc.}$$

The coefficients of the unknown quantities are symmetrical, the first horizontal and vertical rows being alike, the normal equation for s being distinguished by the presence of Σa^2, that for t by Σb^2 and that for u by Σc^2. Now let us proceed to find the value of u from these equations. The value of s from the first equation is

(120)
$$s = \frac{\Sigma aM}{\Sigma a^2} - \frac{\Sigma ab}{\Sigma a^2} . t - \frac{\Sigma ac}{\Sigma a^2} . u.$$

Placing this value of s in the second and third equations, they become

(121)
$$\Sigma_1 b^2 . t + \Sigma_1 bc . u = \Sigma_1 bM,$$
$$\Sigma_1 bc . t + \Sigma_1 c^2 . u = \Sigma_1 cM,$$

provided that we place,

(122)
$$\Sigma_1 b^2 = \Sigma b^2 - \frac{\Sigma ab}{\Sigma a^2} \Sigma ab,$$
$$\Sigma_1 bc = \Sigma bc - \frac{\Sigma ab}{\Sigma a^2} \Sigma bc,$$
$$\Sigma_1 c^2 = \Sigma c^2 - \frac{\Sigma ac}{\Sigma a^2} \Sigma ab,$$
$$\Sigma_1 bM = \Sigma bM - \frac{\Sigma aM}{\Sigma a^2} \Sigma bc,$$
$$\Sigma_1 cm = \Sigma cM - \frac{\Sigma aM}{\Sigma a^2} \Sigma ac.$$

The two equations (121) are then exactly similar in form to the second and third original normal equations, except that the terms containing s have disappeared, and each coefficient is marked with the index 1. From the first of these we take

$$(123) \qquad t = \frac{\Sigma_1 bM}{\Sigma_1 b^2} - \frac{\Sigma_1 bc}{\Sigma_1 b^2} u,$$

and substitute it in the second, giving

$$(124) \qquad \Sigma_2 c^2 . u = \Sigma_2 cM,$$

in which, as before, we preserve the symmetry of the coefficients by writing

$$(125) \qquad \Sigma_2 c^2 = \Sigma_1 c^2 - \frac{\Sigma_1 bc}{\Sigma_1 c^2} \Sigma_1 bc,$$

$$\Sigma_2 cM = \Sigma_1 cM - \frac{\Sigma_1 bM}{\Sigma_1 c^2} \Sigma_1 bc,$$

and hence we have for u,

$$(124) \qquad u = \frac{\Sigma_2 cM}{\Sigma_2 c^2}.$$

Inserting this in (123) we have the value of t, and then from (120) the value of s.

The expressions (125) and (122) which result from the abridged notation are called *auxiliaries*. To deal with equations by this method, we may as Art. 37 form the sums (72) and the normal equations (73), then having obtained the coefficients Σa^2, etc. we insert them in (122) and have the first auxiliaries $\Sigma_1 a^2$, $\Sigma_1 ab$, etc., which being substituted in (125) give the second auxiliaries $\Sigma_2 c^2$, etc., and lastly, from (124), (123) and (120) the values of u, t and s. By the process as thus followed the work is reduced to a routine, which may be followed by one ignorant of the theory of the method. As a check upon the work, the normal equation for s may be taken as the final one and a new set of auxiliaries written, and the values deduced in the reverse order s, t, u. Other checks upon the accuracy of the work are also afforded by the properties of the normal equations, which will readily occur to all computers. In common calculations, however, the use of such routines is not to be recommended.

61. In Art. 39 we have shown that the weight of any unknown quantity as u is the reciprocal of the absolute term in the normal equation for u contained in the general value of u. Now in solving the normal equations as above, it is evident that the absolute term in the normal equation for u is unaffected by substitutions of t and s from the other equations. Hence in the above example $\Sigma_2 c^2$ is the weight of the value of u.

In this manner, the values of the unknown quantities and their weights may be found, and thus the results obtained by the process of Art. 39 be checked by an independent method. In fact, in all calculations where large numbers of observations enter, the solutions should be made by separate methods, and if possible by independent computers.

Other Formulæ for Probable Errors.

62. The formation of the squares of the residuals involves in practice considerable labour, when the residuals are so large that tables of squares cannot be used. Formulæ have hence been deduced for probable errors in which only the residuals themselves are employed. We give here, without demonstration, two such formulæ for the common case of direct observations of equal weight (Arts. 23—26). Let n denote the number of measurements, and Σv the sum of the residuals $v_1, v_2 \ldots v_n$, *all being taken with the positive sign*, then if n is a large number, the probable error of a single observation is, approximately,

(126) $$r = 0 \cdot 8453 \frac{\Sigma v}{n},$$

or more nearly,

(127) $$r = 0 \cdot 8453 \frac{\Sigma v}{\sqrt{n(n-1)}}.$$

Thus in the example of Art. 24 we have 24 observations, and the sum of the residuals all taken positively is $37'''\cdot 48$.

The formula (126) gives $r = 1''\cdot 32$ and (127) gives $r = 1\cdot 344$, while from the stricter formula (42) we found $r = 1\cdot 349$. With a larger value of n, a closer agreement might be expected. The probable error of the arithmetical mean found by dividing r by the square root of n, will in the above example be practically the same by all three formulæ. For values of n less than 24 it is best to hold fast to the more exact formula (42), and even that cannot for such cases be expected to give precise results, since the hypothesis of its development supposes that enough observations have been taken to exhibit the several errors in proportion to their respective probabilities. The formula (127) is due to PETERS.

The Mean Error.

63. The choice of the probable error as a means of comparison of different series of observations is, as we have before mentioned (Art. 55), entirely arbitrary, although it seems to be the most natural one from its middle position in the series of errors (Art. 16). Another error very commonly employed for the same purpose is called the *mean error*, whose definition is, *the error whose square is the mean of the squares of all the errors.* Hence, the mean error is the square root of the quantity $\dfrac{\Sigma x^2}{n}$ (59), and is consequently that part of the probable error included under the radical sign. If then ϵ be the mean and r the probable error,

(128) $$\epsilon = \frac{r}{0\cdot 6745} = 1\cdot 4826\, r.$$

To transform then our expressions for probable error into those for mean error, we have only to omit the constant factor $0\cdot 6745$. In Fig. 1. the abscissa OP denotes the probable and OM the mean error. It is a probability of $\frac{1}{2}$, or a wager of 1 to 1 that an error taken at random is less than the *probable error;* it is a probability of $\dfrac{682}{1000}$ or a wager of $2\cdot 14$ to 1 that it is less than the *mean error.*

190   APPENDIX.

In this book we have chosen to employ only the probable error as being the simplest in theory, and the one in most common use. For the sake of uniformity, it is certainly to be desired that the mean error should be discarded and the former only employed.

List of Literature.

64. The following list gives the titles of some of the most important memoirs and books upon the Method of Least Squares and the law of errors of observations. It is intended to include not only the best text-books, but also those works which are of the greatest historical value. The arrangement is chronological.

It would be easy to greatly extend the limits of this list. The titles have in fact been selected from a list of about four hundred which I hope sometime to publish, accompanied by historical and critical notes. But as an aid to the general reader, a selection such as here given will prove of greater value than if the number were increased tenfold.

1. COTES. Estimatio errorum in mixta mathesi ...; *Harmonia Mensurarum* (Cantabridgiae, 1722, 4to.), pp. 2—22.

2. BOSCOVICH. *De littera expeditione per Pontificiam ditionem ad dimetiendos duos meridiani gradus;* Romae, 1755, 4to., pp. xxii, 516.

3. SIMPSON. An Attempt to show the Advantage of taking the Mean of a Number of Observations...; *Miscellaneous Tracts* (London, 1757, 4to.), pp. 64—75.

4. LAPLACE. Déterminer le milieu que l'on doit prendre entre trois observations.... *Mém. Acad. Paris par divers savans [étrangers]*, 1774, Vol. VI. pp. 634—644.

5. BERNOULLI (Daniel). Dijudicatio maxime probabilis observationum discrepantium atque verisimillima inductio inde formanda; *Acta Acad. Petrop.* for 1777, Pt. I. pp. 3—23.

6. LAPLACE. Chap. v. Book III. of *Traité de mécanique céleste*, Paris, 1799, 4to.

---

7. LEGENDRE. *Nouvelles méthodes pour la détermination des orbites des cométes;* Paris, 1805, 4to., pp. viii, 80.

8. ADRIAN. Research concerning the probabilities of the errors which happen in making observations; *Analyst*, 1808, No. IV. pp. 93—109. See *Amer. Jour. Sci.* 1871, Vol. I. p. 412.

9. GAUSS. Determinatio orbitæ observationibus quotcunque quam proximæ satisfacientis; *Theoria motus corporum cœlestium* (Hamburgi, 1809, 4to.), Lib. II. Sect. III. pp. 205—224.

10. LAPLACE. *Théorie analytique des Probabilités;* Paris, 1812, 4to., pp. 464. Third ed. with supplements, 1820, pp. cxlii, 506, 34, 50, 36.

11. GAUSS. Bestimmung der Genauigkeit der Beobachtungen; *Zeitschr. f. Astr.*, 1816, Vol. I. pp. 185—197.

12. PAUCKER. *Die Anwendung der Methode der kleinsten Quadratsummen auf physikalische Beobachtungen:* Mitau, 1819, 4to., pp. 32.

13. GAUSS. Theoria combinationis observationum erroribus minimis obnoxiæ; *Comment. Soc. Götting.*, 1819—22, Vol. V. pp. 33—90.

14. IVORY. On the Method of the Least Squares: *Phil. Mag.*, 1825, Vol. LXV. pp. 3—10, 81, 161—168; 1826, Vol. LXVIII. pp. 161—165.

15. GAUSS. Supplementum theoriæ combinationis observationum...; *Comment. Soc. Götting.*, 1823—27, Vol. VI. pp. 57—98.

16. POISSON. Mémoire sur la probabilité des résultats moyens des observations; *Connais. des Temps*, 1827, pp. 273—302; 1832, pp. 3—22.

17. ENCKE. Ueber die Begründung der Methode der kleinsten Quadrate; *Abhandl. Acad. Berlin*, 1831, pp. 73—78.

18. ENCKE. Ueber die Methode der kleinsten Quadrate; *Berliner Jahrbuch*, 1834, pp. 249—312; 1835, pp. 253—320; 1836, pp. 253—308.

19. HAGEN. *Grundzüge der Wahrscheinlichkeitsrechnung;* Berlin, 1837, 8vo. Second ed. in 1867, pp. x, 187.

20. BESSEL. Untersuchungen über die Wahrscheinlichkeit der Beobachtungsfehler; *Astr. Nachr.*, 1838, Vol. XV. col. 369—404.

21. BESSEL and BAEYER. *Gradmessung in Ost-Preussen...;* Berlin, 1838, 4to. pp. xiv, 452.

22. GERLING. *Die Ausgleichungsrechnung der practischen Geometrie...;* Hamburg, 1843, 8vo., pp. xix, 409.

23. ELLIS. On the Method of Least Squares; *Trans. Cam. Phil. Soc.;* 1844, Vol. VIII. pp. 204—219.

24. GALLOWAY. On the application of the Method of Least Squares to ... a Portion of the Ordnance Survey of England; *Mem. Astr. Soc. Lond.*, 1846, Vol. XV. pp. 23—69.

25. QUETELET. *Lettres ... sur la Théorie des Probabilités;* Bruxelles, 1846, 8vo., pp. iv, 450.

26. HERSCHEL (J. F. W.). QUETELET on Probabilities. *Edinb. Rev.*, 1850, Vol. XCII. pp. 1—57.

27. ELLIS. Remarks on an alleged Proof of the Method of Least Squares. *Phil. Mag.*, 1850, Vol. XXXVII. pp. 321—328, 462.

28. VERDAM. *Verhandeling over de methode der kleinste quadraten;* Groningen, 1850, Vol. I. large 4to. pp. xxi, 409.

29. DONKIN. Sur la théorie de la combinaison des observations; *Jour. de Mathém.* 1850, Vol. XV. pp. 297—322.

30. BIENAYME. Sur la probabilité des erreurs d'après la méthode des moindres carrés; *Jour. de Math.* 1852. Vol. XVII. pp. 33—78.

31. LIAGRE. *Calcul des probabilités et la théorie des erreurs;* Bruxelles, 1852, 8vo.

32. BERTRAND. *Méthode des moindres carrés.* [Translations of GAUSS' memoirs]; Paris, 1855, 8vo. pp. 167.

33. SCHOTT. Adjustment of horizontal angles of a triangulation... etc. *U.S. Coast Survey Rep.*, 1854, pp. 70—95.

34. DIENGER. *Ausgleichung der Beobachtungsfehler*

APPENDIX.

nach der Methode der kleinsten Quadratsummen; Braunschweig, 1857, 8vo. pp. vii, 168.

✗ 35. RITTER. *Manuel théorique et practique de l'application de la méthode des moindres carrés :* Paris, 1858, 8vo.

36. AIRY. *On the Algebraical and Numerical Theory of Errors of Observation* ...; Cambridge, 1861, 8vo. pp. xvi, 103.—Second Edition in 1875.

37. FREEDEN. *Die Praxis der Methode der kleinsten Quadrate :* Braunschweig, 1863. 8vo. pp. viii, 114.

✗ 38. CHAUVENET. Method of Least Squares. Appendix to his Astronomy (Philadelphia, 1864, 8vo.), Vol. II. pp. 464—566. Also as separate reprint.

.39. DE MORGAN. On the Theory of Errors of Observations; *Trans. Camb. Phil. Soc.,* 1864, Vol. X. pp. 409—427.

40. TODHUNTER. *A History of the mathematical Theory of Probability from the time of Pascal to that of Laplace;* Cambridge and London, 1865, 8vo. pp. xvi, 624.

41. MÜLLER-HAUENFELS. *Höhere Markscheidekunst;* Wien, 1868, 8vo. pp. xii, 291.

42. HANSEN. *Anwendung der Methode der kleinsten Quadraten auf Geodäsie :* Leipzig, 1868, 8vo. pp. 236.

43. TODHUNTER. On the Method of Least Squares; *Trans. Camb. Phil. Soc.,* 1869, Vol. XI. pp. 219—238.

44. CROFTON. On the Proof of the Law of Errors of Observations; *Phil. Trans.,* 1870, pp. 175—188.

45. HELMERT. *Ausgleichungsrechnung nach der Methode der kleinsten Quadrate :* Leipzig, 1872, 8vo. pp. xi, 348.

46. GLAISHER. On the Law of Facility of Errors of Observations, and on the Method of Least Squares; *Mem. ...Astr. Soc. Lond.,* 1872, Vol. XXXIX. Pt. II. pp. 75—124.

47. GLAISHER. On the Rejection of Discordant Observations; *Month. Not....Astr. Soc. Lond.,* 1873, Vol. XXXIII. pp. 392—402.

Faà de Bruno.

On the History of the Method of Least Squares.

65. In the following brief sketch constant reference is understood to be made to the preceding list of literature.

The average or arithmetical mean has always from the earliest times been employed for the combination of direct observations of equal precision made upon a single quantity. Out of this arises so naturally the idea of weights and of the general mean (Art. 27), that it is probable that both were in extensive use long before any attempt to deduce general rules based upon scientific principles. The first recorded discussion of indirect observations with a view of establishing a general method of adjustment seems to be that of COTES, who about the year 1714 introduced the idea of observation equations, or, as they are often called, equations of condition. He took up the simple case of determining a quantity z from measurements upon the related quantity $M = az$, and representing each measurement by an equation, obtained $a_1 z - M_1 = 0$, $a_2 z - M_2 = 0$, etc. Without here entering into the details of his reasoning, we may say that his method consisted essentially in adding the several equations, and from their sum finding the value of z, which value he regarded as the most advantageous or plausible. This process involves the principle that the algebraic sum of the errors shall be zero, and since this coincided with the practice of the arithmetical mean, when the constants a_1, a_2, etc. were equal, it seemed to be the proper adjustment of such cases. It afforded, however, no means for the combination of equations containing more than one unknown quantity. The next step appears to have been made by Pater BOSCOVICH, who being deputed by the Pope to measure an arc of the meridian, published his results in 1755, and who introduced the plan of adjusting observations by introducing, in addition to the principle of COTES, the condition that the sum of the errors *all taken positively* should be a minimum. These two principles rendered possible the adjustment of indirect observations involving several unknown quantities, and appears to have been considerably employed by subsequent writers, particularly by LAPLACE, who in 1799 gave an exhaustive discussion of observations upon the length of the second's pendulum and the resulting ellipticity of the earth. This method, although incorrect, served to prepare the way for a better one.

The first published application of the method of Least Squares is due to LEGENDRE, who in 1806 stated, without demonstration, the principle that the sum of the squares of the errors should be a minimum, and gave an example illustrating its use in determining the orbits of comets. The honour of its discovery and introduction is however universally conceded to GAUSS, who as early as 1795 had used the method in his computations, and had communicated it to his astronomical friends, and who in his *Theoria motus corporum*, published in 1809, gave first its development and demonstration. This proof has been followed by the great majority of subsequent writers, and being the best adapted for an elementary presentation of the subject is also used in this book. To GAUSS is also due the development of the algorithm of the method, the formulæ for probable error, the determination of weights, the method of correlatives, and many other features of the subject, as well as numerous practical applications with which his writings abound. Very few branches of science owe so large a proportion of subject matter to the labours of one man. In 1812 LAPLACE in his treatise on Probabilities took up the subject, and gave an entirely different demonstration of the fundamental principle that the sum of the squares of the errors should be a minimum, and in 1826 GAUSS in his *Supplementum*, etc. again returned to the subject with a second demonstration.

The method thus thoroughly established spread among astronomers with wonderful rapidity. The theory was subjected to rigid analysis and discussion by IVORY, POISSON, ENCKE, and others, while the labours of HANSEN, BESSEL, and GERLING developed its practical applications to astronomical and geodetical observations. The works of BESSEL may be particularly mentioned as establishing the processes since universally employed in extensive trigonometrical surveys. Our partial list of literature indicates the thoroughness with which the subject has been treated by German and French scientists. In this connection we may be allowed to mention as a guide to the student, that Nos. 14, 23, 27 and 46 in that list give interesting critical discussions of the theory of the subject, and that No. 25 is a popular and interesting presentation of the law of probability of error. The Method of Least Squares is now universally employed in all branches of physical science where accurate observations arise, and it is perhaps not too much to say that the precision of astronomical tables is due in great part to its use.

Remarks on the Theory of Least Squares.

66. The proof which we have given of the Method of Least Squares is the one presented by GAUSS in 1809 and followed by the great majority of subsequent writers. It first establishes the law of the probability of error represented by the equation (12) $y = ce^{-h^2x^2}$, from which the principle of the method immediately follows. The whole reasoning is thus dependent upon the theory of probability and upon the particular law of error $y = ce^{-h^2x^2}$. In the demonstration of this law of error in Art. 11b, there are two defects, which I now proceed to point out.

The first is the assumption that the average or arithmetical mean furnishes, for direct observations of equal weight, the *most probable* value of the quantity sought. In the strict mathematical sense of the words "most probable," as defined in Art. 9, this is not true, or, if true, has never yet been proved. In the common sense of the words the arithmetical mean may perhaps be regarded as most correct, most advantageous, most plausible, or even as most probable, but I am unaware that it has ever been shown that the *a priori mathematical probability* of the average is the greatest out of all the probabilities of all the assignable values. Hence in equation (8) it is not really known that the letter z represents the same quantity on both sides of the sign of equality.

The second is the transition from equation (8) to equation (9). Granting that in (8) the value of z is the same in both members, it follows that the corresponding terms are equal each to each. But the quantities $(z - M_1)(z - M_2)$, etc. in that equation are not *actual errors*, since z is not here the *true* but the probable value of the measured quantity. Hence equation (9), in which x represents the actual true error, does not strictly follow from (8) unless we grant that the same law of error which occurs in the particular case (4) obtains also in the general case (3), that is, unless the law which is true for the *residual* is true also for the error.

These two difficulties have puzzled mathematical students since the year 1809, and they cannot be bridged over or avoided, but will always exist in this mathematical development of the law of probability of error. And I think for this reason; it is distinctly conceivable and hence *a priori* possible that in different classes of observations different laws of error might exist, therefore we ought not to expect that from *merely theoretical* considerations a single definite law of error should result. And so we introduce an element derived from *experience*, viz. the arithmetical mean, which we know in a large number of observations gives a very near approximation to the measured quantity, and which in an infinite number of observations would give us its true value. At the limit then when n is infinite, all the equations of Art. 11 are in strict agreement, and one law of error obtains.

The demonstration of LAPLACE given in his *Théorie analytique des Probabilités* employs very different reasoning. It shows that the sum of the squares of the errors must be made a minimum in order to furnish the most advantageous (not the most probable) values of the quantities sought, without any reference to the law of probability of error. The proof is stated by him and by other writers to be entirely general, whatever be the law of probability of error. This point I do not propose to discuss, but will mention as a curious circumstance that AIRY (No. 36 of our list of literature), who follows that method, finds no difficulty in deducing an expression which agrees in every respect with our equation (19), $y = h i \pi^{-\frac{1}{2}} e^{-h^2 x^2}$. I may say also that LAPLACE'S reasoning supposes the number of observations to be infinite. This demonstration is followed as far as I am aware by only three authors, viz. by POISSON, by DE MORGAN and by AIRY; it is severely attacked by IVORY, who proves conclusively, as he says, that it involves the particular law of error deduced by GAUSS' method, and it is defended by ELLIS, who however does not show that IVORY was wrong.

IVORY himself gives no less than three demonstrations of the method of least squares, which are entirely independent of any idea of probability, and which although untenable, furnish extremely interesting reading and deserve to rank high as mathematical curiosities.

A critical and exhaustive discussion of these and other proofs of the Method of Least Squares is given by GLAISHER in a memoir published in 1872. (See No. 46 of the above list of literature.)

In conclusion, I may say that another difficulty sometimes found in GAUSS' proof does not exist as here presented. In the equation (19) $y = hi\pi^{-\frac{1}{2}} e^{-h^2 x^2}$, the probability y is an infinitesimal if x is a continuous variable, owing to the presence of the letter i, and is an abstract number as it ought to be (see Art. 12). The law as given by most writers is $y = h\pi^{-\frac{1}{2}} e^{-h^2 x^2}$, which is absurd, since y is then not only a concrete number but finite, and hence the probability of an error lying between any two assigned limits must be infinity, a conclusion which those authors take care to avoid by an ingenious but questionable artifice.

ALPHABETICAL INDEX.

THE NUMBERS INDICATE PAGES.

Accidental errors, 3, 120
Adjustment, see Observations, etc.

Angle, adjustment of, 29, 36, 38, 56
Angles around a point, 51, 57, 86
  ,,   of a triangle, 73, 78, 88, 99
  ,,   of a quadrilateral, 76, 92
Arithmetical mean, 17, 25, 127, 142, 194
  ,,   probable error of, 26, 147
Axioms, 9, 124

Base line, 33, 39, 63, 93
Bessel's observations, etc., 13, 195
Borings at Grenelle, 102
  ,,   at Speremberg, 104
Boscovich, historical notice of, 194
Bowditch's formula for pendulum, 109.

Certainty, 4, 121
Chemical analysis adjusted, 93
Chronometer, rate of, 65
Cleveland, Ohio, elevation of, 48, 62, 63
Coast Survey Reports, 27, 85, 108
Coins, throwing of, 3, 5, 6, 123
Columbus, Ohio, elevation of, 48
Compass bearings, 26, 31
Compass, declination, 106
Conditional observations, 22, 68, 172
Constant errors, 3, 115, 120
Correlatives, method of, 73, 88, 173
Cotes, historical notice of, 194
Criterion for doubtful observations, 117, 196
Curve of Probability, 11, 125

Declination, magnetic, 106
Direct observations, 22, 24, 41, 141

Earth, temperature of, 102

Empirical formulæ, 100, 179
Equations, conditional, 68, 172
  ,,   correlative, 73, 88, 173
  ,,   normal, 44, 74, 156, 161, 185
  ,,   observation, 41, 153
  ,,   of prob. curve, 10, 131
Errors, 9, 15, 26, 120
  ,,   accidental, 3, 120
  ,,   constant, 2, 115, 120
  ,,   mean, 189
  ,,   probability of, 11, 110, 124, 181
  ,,   see Probable Error

Foot, English and U.S., 110 n.
Formulæ empirical, 100, 179
Friction, coefficient of, 66
Function, defined, 9

Gardner's report on levels, 47
Gauss' law of errors, 10, 126
  ,,   method for weights, 59
  ,,   ,,   for correlatives, 73, 88, 173
  ,,   historical notice of, 195
General mean, 31, 148
  ,,   prob. error of, 36, 149
Gravity at New York, 101
Guessing, problem on, 116

Harrisburg, Penn. elevation of, 48
Hartford, Conn. magnetic declination, 107
History of Least Squares, 194

Independent observations, 22, 46, 153
Indeterminate multipliers, 157
Indirect observations, 22, 46, 153, 194
Instrumental errors, 3

ALPHABETICAL INDEX.

Integrals determined, 133 n., 135 n.
Iron, Cast, coefficient of friction, 66
„ Pig, analysis of, 93
IVORY, on pendulum, 110
„ on theory of least squares, 198

Lake Erie, elevation of, 48
LAPLACE, historical notice, 194, 197
Latitude, determination of, 65
Law of gravity, 101
Law of probability of errors, 8, 124, 181
Laws, discovery of, 100
Least Squares, critical remarks, 196
„ history of, 194
„ method of, 14
„ principle of, 17, 136
„ literature of, 190
Levels, adjustment of, 24, 41, 47, 55, 57, 64, 66, 94
Linear equations, 118, 183
Lines, see Base line
Literature, list of, 190.

Magnetic declination, 106
Mean, arithmetical, 17, 25, 141
„ general, 31, 148
Mean error, 189
Measure of Precision, 10, 19, 131, 138
Meter, length of, 102 n., 110
Mistakes, 2, 120
"Most Probable," 1, 7, 123

Normal Equations, 44, 46, 156, 161, 185

Observation defined, 1
„ equations, 41, 42, 45, 153
Observations, adjustment of, 14, 46
„ comparison of, 17
„ conditioned, 22, 24, 70, 172
„ direct, 22, 24, 41, 141
„ discussion of, 100, 177
„ doubtful, 116
„ errors of, 2, 120
„ independent, 22, 153
„ indirect, 22, 153, 194
„ kinds of, 22, 141

PEIRCE's Criterion, 117
Pendulum, second's, 109
Personal equation, 2
Pittsburgh, Penn. elevation of, 48, 62
Precision, measure of, 10, 18, 131, 138
Probable error, 17, 138
„ of arith. mean, 29, 146, 171
„ of general mean, 36, 149
„ of indirect observations, 60, 163
„ of conditioned observations, 18, 139
„ of single observations, 26, 142, 188
„ of functions, etc. 64, 169
Probability, defined, 4, 121
„ curve of, 8, 10, 125
Probability of errors, law of, 7, 124, 181
„ table of, 12, 112
Probability, problems on, 7

Quadrilateral, adjusted, 80, 85, 92

Residuals, 27, 142

SCHOTT, on magnetic declination, 106, 108
Sets of observations, 36, 148

Temperature of Earth, 102
Triangles, plane, 73, 78, 88, 99
„ spherical, 92
Triangulation, 84, 89

United States *Coast Survey Reports*, 27, 85, 108
United States and English feet, 110 n.

Water, volume and temperature of, 105
Weights, 30, 148
Weights of averages of sets, 25, 148
„ of indirect observations, 60, 166

January, 1877.

A CATALOGUE of EDUCATIONAL BOOKS, Published by MACMILLAN and Co., Bedford Street, Strand, London.

CLASSICAL.

Æschylus.—THE EUMENIDES. The Greek Text, with Introduction, English Notes, and Verse Translation. By BERNARD DRAKE, M.A., late Fellow of King's College, Cambridge. 8vo. 3s. 6d.

Aristotle.— AN INTRODUCTION TO ARISTOTLE'S RHETORIC. With Analysis, Notes, and Appendices. By E. M. COPE, Fellow and Tutor of Trinity Coll. Cambridge. 8vo. 14s.

ARISTOTLE ON FALLACIES; OR, THE SOPHISTICI ELENCHI. With Translation and Notes by E. POSTE, M.A., Fellow of Oriel College, Oxford. 8vo. 8s. 6d.

Aristophanes.—THE BIRDS. Translated into English Verse, with Introduction, Notes, and Appendices, by B. H. KENNEDY, D.D., Regius Professor of Greek in the University of Cambridge. Crown 8vo. 6s.

Belcher.—SHORT EXERCISES IN LATIN PROSE COMPOSITION AND EXAMINATION PAPERS IN LATIN GRAMMAR, to which is prefixed a Chapter on Analysis of Sentences. By the Rev. H. BELCHER, M.A., Assistant Master in King's College School, London. 18mo. 1s. 6d. Key, 1s. 6d.

Blackie.—GREEK AND ENGLISH DIALOGUES FOR USE IN SCHOOLS AND COLLEGES. By JOHN STUART BLACKIE, Professor of Greek in the Univ. of Edinburgh. Second Edition. Fcap. 8vo. 2s. 6d.

Cicero. — THE SECOND PHILIPPIC ORATION. With Introduction and Notes. From the German of KARL HALM. Edited, with Corrections and Additions, by Professor JOHN E. B. MAYOR, M.A., Fellow and Classical Lecturer of St. John's College, Cambridge. Fourth Edition, revised. Fcap. 8vo. 5s.

THE ORATIONS OF CICERO AGAINST CATILINA. With Notes and an Introduction. From the German of KARL HALM, with additions by Professor A. S. WILKINS, M.A., Owens College, Manchester. New Edition. Fcap. 8vo. 3s. 6d.

A

Cicero—THE ACADEMICA OF CICERO. The Text revised and explained by JAMES REID, M.A., Assistant Tutor and late Fellow of Christ's College, Cambridge. Fcap. 8vo. 4s. 6d.

Demosthenes.—ON THE CROWN, to which is prefixed ÆSCHINES AGAINST CTESIPHON. The Greek Text with English Notes. By B. DRAKE, M.A., late Fellow of King's College, Cambridge. Fifth Edition. Fcap. 8vo. 5s.

Ellis.—PRACTICAL HINTS ON THE QUANTITATIVE PRONUNCIATION OF LATIN, for the use of Classical Teachers and Linguists. By A. J. ELLIS, B.A., F.R.S. Extra fcap. 8vo. 4s. 6d.

Goodwin.—SYNTAX OF THE MOODS AND TENSES OF THE GREEK VERB. By W. W. GOODWIN, Ph.D. New Edition, revised. Crown 8vo. 6s. 6d.
"*This scholarly and exhaustive work.*"—SCHOOL BOARD CHRONICLE.

Greenwood.—THE ELEMENTS OF GREEK GRAMMAR, including Accidence, Irregular Verbs, and Principles of Derivation and Composition; adapted to the System of Crude Forms. By J. G. GREENWOOD, Principal of Owens College, Manchester. Fifth Edition. Crown 8vo. 5s. 6d.

Hodgson.—MYTHOLOGY FOR LATIN VERSIFICATION. A brief Sketch of the Fables of the Ancients, prepared to be rendered into Latin Verse for Schools. By F. HODGSON, B.D., late Provost of Eton. New Edition, revised by F. C. HODGSON, M.A. 18mo. 3s.

Homer's Odyssey.—THE NARRATIVE OF ODYSSEUS. With a Commentary by JOHN E. B. MAYOR, M.A., Kennedy Professor of Latin at Cambridge. Part I. Book IX.—XII. Fcap. 8vo. 3s.

Horace.—THE WORKS OF HORACE, rendered into English Prose, with Introductions, Running Analysis, and Notes, by J. LONSDALE, M.A., and S. LEE, M.A. Globe 8vo. 3s. 6d.
THE ODES OF HORACE IN A METRICAL PARAPHRASE. By R. M. HOVENDEN, B.A., formerly of Trinity College, Cambridge. Extra fcap. 8vo. 4s. 6d.

Jackson.—FIRST STEPS TO GREEK PROSE COMPOSITION. By BLOMFIELD JACKSON, M.A. Assistant-Master in King's College School, London. Second Edition, revised and enlarged. 18mo. 1s. 6d.
"*A capital little book for beginners.*"—SPECTATOR.

Juvenal.—THIRTEEN SATIRES OF JUVENAL. With a Commentary. By JOHN E. B. MAYOR, M.A., Kennedy Professor of Latin at Cambridge. Second Edition, enlarged. Vol. I. Crown 8vo. 7s. 6d. Or Parts I. and II. Crown 8vo. 3s. 6d. each.

CLASSICAL.

Marshall.—A TABLE OF IRREGULAR GREEK VERBS, classified according to the arrangement of Curtius' Greek Grammar. By J. M. MARSHALL, M.A., Fellow and late Lecturer of Brasenose College, Oxford; one of the Masters in Clifton College. 8vo. cloth. New Edition. 1s.

Mayor (John E. B.)—FIRST GREEK READER. Edited after KARL HALM, with Corrections and large Additions by Professor JOHN E. B. MAYOR, M.A., Fellow and Classical Lecturer of St. John's College, Cambridge. New Edition, revised. Fcap. 8vo. 4s. 6d.

BIBLIOGRAPHICAL CLUE TO LATIN LITERATURE. Edited after HÜBNER, with Large Additions by Professor JOHN E. B. MAYOR. Crown 8vo. 6s. 6d.

"*An extremely useful volume that should be in the hands of all scholars.*"—ATHENÆUM.

Mayor (Joseph B.)—GREEK FOR BEGINNERS. By the Rev. J. B. MAYOR, M.A., Professor of Classical Literature in King's College, London. Part I., with Vocabulary, 1s. 6d. Parts II. and III., with Vocabulary and Index, 3s. 6d., complete in one vol. New Edition. Fcap. 8vo. cloth, 4s. 6d.

Nixon.—PARALLEL EXTRACTS arranged for translation into English and Latin, with Notes on Idioms. By J. E. NIXON, M.A., Classical Lecturer, King's College, London. Part I.—Historical and Epistolary. Crown 8vo. 3s. 6d.

A FEW NOTES ON LATIN RHETORIC. With Tables and Illustrations. By J. E. NIXON, M.A. Crown 8vo. 2s.

Peile (John, M.A.)—AN INTRODUCTION TO GREEK AND LATIN ETYMOLOGY. By JOHN PEILE, M.A., Fellow and Tutor of Christ's College, Cambridge, formerly Teacher of Sanskrit in the University of Cambridge. Third and Revised Edition. Crown 8vo. 10s. 6d.

"*A very valuable contribution to the science of language.*"—SATURDAY REVIEW.

Plato.—THE REPUBLIC OF PLATO, Translated into English, with an Analysis and Notes, by J. LL. DAVIES, M.A., and D. J. VAUGHAN, M.A. Third Edition, with Vignette Portraits of Plate and Socrates, engraved by JEENS from an Antique Gem. 18mo. 4s. 6d.

Plautus.—THE MOSTELLARIA OF PLAUTUS. With Notes Prolegomena, and Excursus. By WILLIAM RAMSAY, M.A., formerly Professor of Humanity in the University of Glasgow. Edited by Professor GEORGE G. RAMSAY, M.A., of the University of Glasgow. 8vo. 14s.

A 2

Potts, Alex. W., M.A.—HINTS TOWARDS LATIN PROSE COMPOSITION. By ALEX. W. POTTS, M.A., LL.D., late Fellow of St. John's College, Cambridge; Assistant Master in Rugby School; and Head Master of the Fettes College, Edinburgh. New Edition, enlarged. Extra fcap. 8vo. cloth. 3*s*.

Roby.—A GRAMMAR OF THE LATIN LANGUAGE, from Plautus to Suetonius. By H. J. ROBY, M.A., late Fellow of St. John's College, Cambridge. In Two Parts. Second Edition. Part I. containing :—Book I. Sounds. Book II. Inflexions. Book III. Word-formation. Appendices. Crown 8vo. 8*s*. 6*d*. Part II.—Syntax, Prepositions, &c. Crown 8vo. 10*s*. 6*d*.
"*Marked by the clear and practised insight of a master in his art. A book that would do honour to any country.*"—ATHENÆUM.

Rust.—FIRST STEPS TO LATIN PROSE COMPOSITION. By the Rev. G. RUST, M.A. of Pembroke College, Oxford, Master of the Lower School, King's College, London. New Edition. 18mo. 1*s*. 6*d*.

Sallust.—CAII SALLUSTII CRISPI CATILINA ET JUGURTHA. For Use in Schools. With copious Notes. By C. MERIVALE, B.D. New Edition, carefully revised and enlarged. Fcap. 8vo. 4*s*. 6*d*. Or separately, 2*s*. 6*d*. each.

Tacitus.—THE HISTORY OF TACITUS TRANSLATED INTO ENGLISH. By A. J. CHURCH, M.A., and W. J. BRODRIBB, M.A. With Notes and a Map. New and Cheaper Edition. Crown 8vo. 6*s*.
"*A scholarly and faithful translation.*"—SPECTATOR.

TACITUS, THE AGRICOLA AND GERMANIA OF. A Revised Text, English Notes, and Maps. By A. J. CHURCH, M.A., and W. J. BRODRIBB, M.A. New Edition. Fcap. 8vo. 3*s*. 6*d*. Or separately, 2*s*. each.
"*'A model of careful editing,' being at once compact, complete, and correct, as well as neatly printed and elegant in style.*"—ATHENÆUM.

THE AGRICOLA AND GERMANIA. Translated into English by A. J. CHURCH, M.A., and W. J. BRODRIBB, M.A. With Maps and Notes. Crown 8vo. New Edition in the press.

TACITUS.—THE ANNALS. Translated, with Notes and Maps, by A. J. CHURCH and W. J. BRODRIBB. Crown 8vo. 7*s*. 6*d*.

Theophrastus.—THE CHARACTERS OF THEOPHRASTUS. An English Translation from a Revised Text.

With Introduction and Notes. By R. C. JEBB, M.A., Professor of Greek in the University of Glasgow. Extra fcap. 8vo. 6s. 6d.

"*A very handy and scholarly edition.*"—SATURDAY REVIEW.

Thring.—Works by the Rev. E. THRING, M.A., Head Master of Uppingham School.

A LATIN GRADUAL. A First Latin Construing Book for Beginners. New Edition, enlarged, with Coloured Sentence Maps. Fcap. 8vo. 2s. 6d.

A MANUAL OF MOOD CONSTRUCTIONS. Fcap. 8vo. 1s. 6d.

A CONSTRUING BOOK. Fcap. 8vo. 2s. 6d.

Thucydides.—THE SICILIAN EXPEDITION. Being Books VI. and VII. of Thucydides, with Notes. New Edition, revised and enlarged, with Map. By the Rev. PERCIVAL FROST, M.A. Fcap. 8vo. 5s.

Virgil.—THE WORKS OF VIRGIL RENDERED INTO ENGLISH PROSE, with Notes, Introductions, Running Analysis, and an Index, by JAMES LONSDALE, M.A. and SAMUEL LEE, M.A. Second Edition. Globe 8vo. 3s. 6d.; gilt edges, 4s. 6d.

"*A more complete edition of Virgil in English it is scarcely possible to conceive than the scholarly work before us.*"—GLOBE.

Wright.—Works by J. WRIGHT, M.A., late Head Master of Sutton Coldfield School.

HELLENICA; OR, A HISTORY OF GREECE IN GREEK, as related by Diodorus and Thucydides; being a First Greek Reading Book, with explanatory Notes, Critical and Historical. Third Edition, with a Vocabulary. 12mo. 3s. 6d.

A HELP TO LATIN GRAMMAR; or, The Form and Use of Words in Latin, with Progressive Exercises. Crown 8vo. 4s. 6d.

THE SEVEN KINGS OF ROME. An Easy Narrative, abridged from the First Book of Livy by the omission of Difficult Passages; being a First Latin Reading Book, with Grammatical Notes. Fifth Edition. With Vocabulary, 3s. 6d.

FIRST LATIN STEPS; OR, AN INTRODUCTION BY A SERIES OF EXAMPLES TO THE STUDY OF THE LATIN LANGUAGE. Crown 8vo. 5s.

ATTIC PRIMER. Arranged for the Use of Beginners. Extra fcap. 8vo. 4s. 6d.

MATHEMATICS.

Airy.—Works by Sir G. B. AIRY, K.C.B., Astronomer Royal:—
ELEMENTARY TREATISE ON PARTIAL DIFFERENTIAL EQUATIONS. Designed for the Use of Students in the Universities. With Diagrams. New Edition. Crown 8vo. cloth. 5s. 6d.

ON THE ALGEBRAICAL AND NUMERICAL THEORY OF ERRORS OF OBSERVATIONS AND THE COMBINATION OF OBSERVATIONS. New edition, revised. Crown 8vo. cloth. 6s. 6d.

UNDULATORY THEORY OF OPTICS. Designed for the Use of Students in the University. New Edition. Crown 8vo. cloth. 6s. 6d.

ON SOUND AND ATMOSPHERIC VIBRATIONS. With the Mathematical Elements of Music. Designed for the Use of Students of the University. Second Edition, Revised and Enlarged. Crown 8vo. 9s.

A TREATISE OF MAGNETISM. Designed for the use of Students in the University. Crown 8vo. 9s. 6d.

Airy (Osmund).—A TREATISE ON GEOMETRICAL OPTICS. Adapted for the use of the Higher Classes in Schools. By OSMUND AIRY, B.A., one of the Mathematical Masters in Wellington College. Extra fcap. 8vo. 3s. 6d.

Bayma.—THE ELEMENTS OF MOLECULAR MECHANICS. By JOSEPH BAYMA, S.J., Professor of Philosophy, Stonyhurst College. Demy 8vo. cloth. 10s. 6d.

Beasley.—AN ELEMENTARY TREATISE ON PLANE TRIGONOMETRY. With Examples. By R. D. BEASLEY, M.A., Head Master of Grantham Grammar School. Fourth Edition, revised and enlarged. Crown 8vo. cloth. 3s. 6d.

Blackburn (Hugh).—ELEMENTS OF PLANE TRIGONOMETRY, for the use of the Junior Class of Mathematics in the University of Glasgow. By HUGH BLACKBURN, M.A., Professor of Mathematics in the University of Glasgow. Globe 8vo. 1s. 6d.

Boole.—Works by G. BOOLE, D.C.L., F.R.S., late Professor of Mathematics in the Queen's University, Ireland.
A TREATISE ON DIFFERENTIAL EQUATIONS. New and Revised Edition. Edited by I. TODHUNTER. Crown 8vo. cloth. 14s.

MATHEMATICS.

Boole—*continued.*

A TREATISE ON DIFFERENTIAL EQUATIONS. Supplementary Volume. Edited by I. TODHUNTER. Crown 8vo. cloth. 8s. 6d.

THE CALCULUS OF FINITE DIFFERENCES. Crown 8vo. cloth. 10s. 6d. New Edition, revised by J. F. MOULTON.

Brook-Smith (J.)—ARITHMETIC IN THEORY AND PRACTICE. By J. BROOK-SMITH, M.A., LL.B., St. John's College, Cambridge; Barrister-at-Law; one of the Masters of Cheltenham College. New Edition, revised. Complete, Crown 8vo. 4s. 6d.
"*A valuable Manual of Arithmetic of the Scientific kind. The best we have seen.*"—LITERARY CHURCHMAN.

Cambridge Senate-House Problems and Riders, WITH SOLUTIONS :—
1848-1851.—RIDERS. By JAMESON. 8vo. cloth. 7s. 6d.
1857.— PROBLEMS AND RIDERS. By CAMPION and WALTON. 8vo. cloth. 8s. 6d.
1864.—PROBLEMS AND RIDERS. By WALTON and WILKINSON. 8vo. cloth. 10s. 6d.
1875.—PROBLEMS AND RIDERS. By A. G. GREENHILL, M.A. Crown 8vo. 8s. 6d.

CAMBRIDGE COURSE OF ELEMENTARY NATURAL PHILOSOPHY, for the Degree of B.A. Originally compiled by J. C. SNOWBALL, M.A., late Fellow of St. John's College. Fifth Edition, revised and enlarged, and adapted for the Middle-Class Examinations by THOMAS LUND, B.D. Crown 8vo. cloth. 5s.

Candler.—HELP TO ARITHMETIC. Designed for the use of Schools. By H. CANDLER, M.A., Mathematical Master of Uppingham School. Extra fcap. 8vo. 2s. 6d.

Cheyne.—Works by C. H. H. CHEYNE, M.A., F.R.A.S.

AN ELEMENTARY TREATISE ON THE PLANETARY THEORY. With a Collection of Problems. Second Edition. Crown 8vo. cloth. 6s. 6d.

THE EARTH'S MOTION OF ROTATION. Crown 8vo. 3s. 6d.

Childe.—THE SINGULAR PROPERTIES OF THE ELLIPSOID AND ASSOCIATED SURFACES OF THE NTH DEGREE. By the Rev. G. F. CHILDE, M.A., Author of "Ray Surfaces," "Related Caustics," &c. 8vo. 10s. 6d.

Christie.—A COLLECTION OF ELEMENTARY TEST-QUESTIONS IN PURE AND MIXED MATHEMATICS; with Answers and Appendices on Synthetic Division, and on the Solution of Numerical Equations by Horner's Method. By JAMES R. CHRISTIE, F.R.S., Royal Military Academy, Woolwich. Crown 8vo. cloth. 8s. 6d.

Cumming.—AN INTRODUCTION TO THE THEORY OF ELECTRICITY. By LINNÆUS CUMMING, M.A. With Illustrations. Crown 8vo. 8s. 6d.

Cuthbertson—EUCLIDIAN GEOMETRY. By FRANCIS CUTHBERTSON, M.A., LL.D., Head Mathematical Master of the City of London School. Extra fcap. 8vo. 4s. 6d.

Dalton.—Works by the Rev. T. DALTON, M.A., Assistant Master of Eton College.

RULES AND EXAMPLES IN ARITHMETIC. New Edition. 18mo. cloth. 2s. 6d. *Answers to the Examples are appended.*

RULES AND EXAMPLES IN ALGEBRA. Part I. 18mo. 2s. Part II. 18mo. 2s. 6d.

Day.—PROPERTIES OF CONIC SECTIONS PROVED GEOMETRICALLY. PART I., THE ELLIPSE, with Problems. By the Rev. H. G. DAY, M.A., Head Master of Sedburgh Grammar School. Crown 8vo. 3s. 6d.

Dodgson.—AN ELEMENTARY TREATISE ON DETERMINANTS, with their Application to Simultaneous Linear Equations and Algebraical Geometry. By CHARLES L. DODGSON, M.A. Small 4to. cloth. 10s. 6d.

Drew.—GEOMETRICAL TREATISE ON CONIC SECTIONS. By W. H. DREW, M.A., St. John's College, Cambridge. Fifth Edition, enlarged. Crown 8vo. cloth. 5s.

SOLUTIONS TO THE PROBLEMS IN DREW'S CONIC SECTIONS. Crown 8vo. cloth. 4s. 6d.

Edgar (J. H.) and Pritchard (G. S.)—NOTE-BOOK ON PRACTICAL SOLID OR DESCRIPTIVE GEOMETRY. Containing Problems with help for Solutions. By J. H. EDGAR, M.A., Lecturer on Mechanical Drawing at the Royal School of Mines, and G. S. PRITCHARD. Third Edition, revised and enlarged. Globe 8vo. 3s.

Ferrers.—AN ELEMENTARY TREATISE ON TRILINEAR CO-ORDINATES, the Method of Reciprocal Polars, and the Theory of Projectors. By the Rev. N. M. FERRERS, M.A., Fellow and Tutor of Gonville and Caius College, Cambridge. Third Edition, revised. Crown 8vo. 6s. 6d.

MATHEMATICS.

Frost.—Works by PERCIVAL FROST, M.A., formerly Fellow of St. John's College, Cambridge; Mathematical Lecturer of King's College.
AN ELEMENTARY TREATISE ON CURVE TRACING. By PERCIVAL FROST, M.A. 8vo. 12s.
THE FIRST THREE SECTIONS OF NEWTON'S PRINCIPIA. With Notes and Illustrations. Also a collection of Problems, principally intended as Examples of Newton's Methods. By PERCIVAL FROST, M.A. Second Edition. 8vo. cloth. 10s. 6d.
SOLID GEOMETRY. A New Edition, revised and enlarged, of the Treatise by FROST and WOLSTENHOLME. In 2 Vols. Vol. I. 8vo. 16s.

Godfray.—Works by HUGH GODFRAY, M.A., Mathematical Lecturer at Pembroke College, Cambridge.
A TREATISE ON ASTRONOMY, for the Use of Colleges and Schools. New Edition. 8vo. cloth. 12s. 6d.
AN ELEMENTARY TREATISE ON THE LUNAR THEORY, with a Brief Sketch of the Problem up to the time of Newton. Second Edition, revised. Crown 8vo. cloth. 5s. 6d.

Hemming.—AN ELEMENTARY TREATISE ON THE DIFFERENTIAL AND INTEGRAL CALCULUS, for the Use of Colleges and Schools. By G. W. HEMMING, M.A., Fellow of St. John's College, Cambridge. Second Edition, with Corrections and Additions. 8vo. cloth. 9s.

Jackson.—GEOMETRICAL CONIC SECTIONS. An Elementary Treatise in which the Conic Sections are defined as the Plane Sections of a Cone, and treated by the Method of Projection. By J. STUART JACKSON, M.A., late Fellow of Gonville and Caius College, Cambridge. 4s. 6d.

Jellet (John H.)—A TREATISE ON THE THEORY OF FRICTION. By JOHN H. JELLET, B.D., Senior Fellow of Trinity College, Dublin; President of the Royal Irish Academy. 8vo. 8s. 6d.

Jones and Cheyne.—ALGEBRAICAL EXERCISES. Progressively arranged. By the Rev. C. A. JONES, M.A., and C. H. CHEYNE, M.A., F.R.A.S., Mathematical Masters of Westminster School. New Edition. 18mo. cloth. 2s. 6d.

Kelland and Tait.—INTRODUCTION TO QUATERNIONS, with numerous examples. By P. KELLAND, M.A., F.R.S.; and P. G. TAIT, M.A., Professors in the department of Mathematics in the University of Edinburgh. Crown 8vo. 7s. 6d.

Kitchener.—A GEOMETRICAL NOTE-BOOK, containing Easy Problems in Geometrical Drawing preparatory to the Study of Geometry. For the Use of Schools. By F. E. KITCHENER, M.A., Mathematical Master at Rugby. Third Edition. 4to. 2s.

Morgan.—A COLLECTION OF PROBLEMS AND EXAMPLES IN MATHEMATICS. With Answers. By H. A. MORGAN, M.A., Sadlerian and Mathematical Lecturer of Jesus College, Cambridge. Crown 8vo. cloth. 6s. 6d.

Newton's PRINCIPIA. Edited by Professor Sir W. THOMSON and Professor BLACKBURN. 4to. cloth. 31s. 6d.
"*Undoubtedly the finest edition of the text of the 'Principia' which has hitherto appeared.*"—EDUCATIONAL TIMES.

Parkinson.—Works by S. PARKINSON, D.D., F.R.S., Tutor and Prælector of St. John's College, Cambridge.
AN ELEMENTARY TREATISE ON MECHANICS. For the Use of the Junior Classes at the University and the Higher Classes in Schools. With a Collection of Examples. Fifth edition, revised. Crown 8vo. cloth. 9s. 6d.
A TREATISE ON OPTICS. Third Edition, revised and enlarged. Crown 8vo. cloth. 10s. 6d.

Phear.—ELEMENTARY HYDROSTATICS. With Numerous Examples. By J. B. PHEAR, M.A., Fellow and late Assistant Tutor of Clare College, Cambridge. Fourth Edition. Crown 8vo. cloth. 5s. 6d.

Pirie.—LESSONS ON RIGID DYNAMICS. By the Rev. G. PIRIE, M.A., Fellow and Tutor of Queen's College, Cambridge. Crown 8vo. 6s.

Pratt.—A TREATISE ON ATTRACTIONS, LAPLACE'S FUNCTIONS, AND THE FIGURE OF THE EARTH. By JOHN H. PRATT, M.A., Archdeacon of Calcutta. Fourth Edition. Crown 8vo. cloth. 6s. 6d.

Puckle.—AN ELEMENTARY TREATISE ON CONIC SECTIONS AND ALGEBRAIC GEOMETRY. With Numerous Examples and Hints for their Solution; especially designed for the Use of Beginners. By G. H. PUCKLE, M.A. New Edition, revised and enlarged. Crown 8vo. cloth. 7s. 6d.

Rawlinson.—ELEMENTARY STATICS, by the Rev. GEORGE RAWLINSON, M.A. Edited by the Rev. EDWARD STURGES, M.A. Crown 8vo. cloth. 4s. 6d.

Reynolds.—MODERN METHODS IN ELEMENTARY GEOMETRY. By E. M. REYNOLDS, M.A., Mathematical Master in Clifton College. Crown 8vo. 3s. 6d.

Routh.—AN ELEMENTARY TREATISE ON THE DYNAMICS OF THE SYSTEM OF RIGID BODIES. With Numerous Examples. By EDWARD JOHN ROUTH, M.A., late Fellow and Assistant Tutor of St. Peter's College, Cambridge; Examiner in the University of London. New and enlarged Edition in the press.

WORKS
By the REV. BARNARD SMITH, M.A.,

Rector of Glaston, Rutland, late Fellow and Senior Bursar of St. Peter's College, Cambridge.

ARITHMETIC AND ALGEBRA, in their Principles and Application; with numerous systematically arranged Examples taken from the Cambridge Examination Papers, with especial reference to the Ordinary Examination for the B.A. Degree. Thirteenth Edition, carefully revised. Crown 8vo. cloth. 10s. 6d.

"*To all those whose minds are sufficiently developed to comprehend the simplest mathematical reasoning, and who have not yet thoroughly mastered the principles of Arithmetic and Algebra, it is calculated to be of great advantage.*"—ATHENÆUM. "*Mr. Smith's work is a most useful publication. The rules are stated with great clearness. The examples are well selected, and worked out with just sufficient detail, without being encumbered by too minute explanations: and there prevails throughout it that just proportion of theory and practice which is the crowning excellence of an elementary work.*" —DEAN PEACOCK.

ARITHMETIC FOR SCHOOLS. New Edition. Crown 8vo. cloth. 4s. 6d. Adapted from the Author's work on "Arithmetic and Algebra."

"*Admirably adapted for instruction, combining just sufficient theory with a large and well-selected collection of exercises for practice.*"—JOURNAL OF EDUCATION.

A KEY TO THE ARITHMETIC FOR SCHOOLS. Tenth Edition. Crown 8vo. cloth. 8s. 6d.

EXERCISES IN ARITHMETIC. With Answers. Crown 8vo. limp cloth. 2s. 6d.
Or sold separately, Part I. 1s.; Part II. 1s.; Answers, 6d.

SCHOOL CLASS-BOOK OF ARITHMETIC. 18mo. cloth. 3s.
Or sold separately, Parts I. and II. 10d. each; Part III. 1s.

KEYS TO SCHOOL CLASS-BOOK OF ARITHMETIC. Complete in one volume, 18mo. cloth, 6s. 6d.; or Parts I., II., and III., 2s. 6d. each.

SHILLING BOOK OF ARITHMETIC FOR NATIONAL AND ELEMENTARY SCHOOLS. 18mo. cloth. Or separately, Part I. 2d.; Part II. 3d.; Part III. 7d. Answers, 6d.

THE SAME, with Answers complete. 18mo. cloth. 1s. 6d.
KEY TO SHILLING BOOK OF ARITHMETIC. 18mo. cloth. 4s. 6d.

EXAMINATION PAPERS IN ARITHMETIC. 18mo. cloth. 1s. 6d. The same, with Answers, 18mo. 1s. 9d.

KEY TO EXAMINATION PAPERS IN ARITHMETIC. 18mo. cloth. 4s. 6d.

Barnard Smith—*continued.*

THE METRIC SYSTEM OF ARITHMETIC, ITS PRINCIPLES AND APPLICATION, with numerous Examples, written expressly for Standard V. in National Schools. Third Edition. 18mo. cloth, sewed. 3*d.*

A CHART OF THE METRIC SYSTEM, on a Sheet, size 42 in. by 34 in. on Roller, 1*s.* 6*d.*, mounted and varnished, price 3*s.* 6*d.* Fourth Edition.

"*We do not remember that ever we have seen teaching by a chart more happily carried out.*"—SCHOOL BOARD CHRONICLE.

Also a Small Chart on a Card, price 1*d.*

EASY LESSONS IN ARITHMETIC, combining Exercises in Reading, Writing, Spelling, and Dictation. Part I. for Standard I. in National Schools. Crown 8vo. 9*d.*

Diagrams for School-room walls in preparation.

"*We should strongly advise everyone to study carefully Mr. Barnard Smith's Lessons in Arithmetic, Writing, and Spelling. A more excellent little work for a first introduction to knowledge cannot well be written. Mr. Smith's larger Text-books on Arithmetic and Algebra are already most favourably known, and he has proved now that the difficulty of writing a text-book which begins ab ovo is really surmountable; but we shall be much mistaken if this little book has not cost its author more thought and mental labour than any of his more elaborate text-books. The plan to combine arithmetical lessons with those in reading and spelling is perfectly novel, and it is worked out in accordance with the aims of our National Schools; and we are convinced that its general introduction in all elementary schools throughout the country will produce great educational advantages.*"—WESTMINSTER REVIEW.

EXAMINATION CARDS IN ARITHMETIC. (Dedicated to Lord Sandon.) With Answers and Hints.
Standards I. and II. in box, 1*s.* 6*d.* Standards III. IV. and V. in boxes, 1*s.* 6*d.* each. Standard VI. in Two Parts, in boxes, 1*s.* 6*d.* each.

A and B papers, of nearly the same difficulty, are given so as to prevent copying, and the Colours of the A and B papers differ in each Standard, and from those of every other Standard, so that a master or mistress can see at a glance whether the children have the proper papers.

Snowball.—THE ELEMENTS OF PLANE AND SPHERICAL TRIGONOMETRY; with the Construction and Use of Tables of Logarithms. By J. C. SNOWBALL, M.A. Eleventh Edition. Crown 8vo. cloth. 7*s.* 6*d.*

SYLLABUS OF PLANE GEOMETRY (corresponding to Euclid, Books I.—VI.) Prepared by the Association for the Improvement of Geometrical Teaching. Second Edition. Crown 8vo. 1*s.*

Tait and Steele.—A TREATISE ON DYNAMICS OF A PARTICLE. With numerous Examples. By Professor TAIT and Mr. STEELE. New Edition, enlarged. Crown 8vo. cloth. 10*s.* 6*d.*

MATHEMATICS.

Tebay.—ELEMENTARY MENSURATION FOR SCHOOLS. With numerous Examples. By SEPTIMUS TEBAY, B.A., Head Master of Queen Elizabeth's Grammar School, Rivington. Extra fcap. 8vo. 3s. 6d.

WORKS

By I. TODHUNTER, M.A., F.R.S.,

Of St. John's College, Cambridge.

"*Mr. Todhunter is chiefly known to students of Mathematics as the author of a series of admirable mathematical text-books, which possess the rare qualities of being clear in style and absolutely free from mistakes, typographical or other.*"—SATURDAY REVIEW.

THE ELEMENTS OF EUCLID. For the Use of Colleges and Schools. New Edition. 18mo. cloth. 3s. 6d.

MENSURATION FOR BEGINNERS. With numerous Examples. New Edition. 18mo. cloth. 2s. 6d.

ALGEBRA FOR BEGINNERS. With numerous Examples. New Edition. 18mo. cloth. 2s. 6d.

KEY TO ALGEBRA FOR BEGINNERS. Crown 8vo. cloth. 6s. 6d.

TRIGONOMETRY FOR BEGINNERS. With numerous Examples. New Edition. 18mo. cloth. 2s. 6d.

KEY TO TRIGONOMETRY FOR BEGINNERS. Crown 8vo. 8s. 6d.

MECHANICS FOR BEGINNERS. With numerous Examples. New Edition. 18mo. cloth. 4s. 6d.

ALGEBRA. For the Use of Colleges and Schools. Seventh Edition, containing two New Chapters and Three Hundred miscellaneous Examples. Crown 8vo. cloth. 7s. 6d.

KEY TO ALGEBRA FOR THE USE OF COLLEGES AND SCHOOLS. Crown 8vo. 10s. 6d.

AN ELEMENTARY TREATISE ON THE THEORY OF EQUATIONS. Third Edition, revised. Crown 8vo. cloth. 7s. 6d.

PLANE TRIGONOMETRY. For Schools and Colleges. Fifth Edition. Crown 8vo. cloth. 5s.

KEY TO PLANE TRIGONOMETRY. Crown 8vo. 10s. 6d.

A TREATISE ON SPHERICAL TRIGONOMETRY. Third Edition, enlarged. Crown 8vo. cloth. 4s. 6d.

PLANE CO-ORDINATE GEOMETRY, as applied to the Straight Line and the Conic Sections. With numerous Examples. Fifth Edition, revised and enlarged. Crown 8vo. cloth. 7s. 6d.

Todhunter (I.)—*continued.*

A TREATISE ON THE DIFFERENTIAL CALCULUS. With numerous Examples. Seventh Edition. Crown 8vo. cloth. 10s. 6d.

A TREATISE ON THE INTEGRAL CALCULUS AND ITS APPLICATIONS. With numerous Examples. Fourth Edition, revised and enlarged. Crown 8vo. cloth. 10s. 6d.

EXAMPLES OF ANALYTICAL GEOMETRY OF THREE DIMENSIONS. Third Edition, revised. Crown 8vo. cloth. 4s.

A TREATISE ON ANALYTICAL STATICS. With numerous Examples. Fourth Edition, revised and enlarged. Crown 8vo. cloth. 10s. 6d.

A HISTORY OF THE MATHEMATICAL THEORY OF PROBABILITY, from the time of Pascal to that of Laplace. 8vo. 18s.

RESEARCHES IN THE CALCULUS OF VARIATIONS, principally on the Theory of Discontinuous Solutions: an Essay to which the Adams Prize was awarded in the University of Cambridge in 1871. 8vo. 6s.

A HISTORY OF THE MATHEMATICAL THEORIES OF ATTRACTION, AND THE FIGURE OF THE EARTH, from the time of Newton to that of Laplace. 2 vols. 8vo. 24s.
"*Such histories are at present more valuable than original work. They at once enable the Mathematician to make himself master of all that has been done on the subject, and also give him a clue to the right method of dealing with the subject in future by showing him the paths by which advance has been made in the past . . . It is with unmingled satisfaction that we see this branch adopted as his special subject by one whose cast of mind and self culture have made him one of the most accurate, as he certainly is the most learned, of Cambridge Mathematicians.*"—SATURDAY REVIEW.

AN ELEMENTARY TREATISE ON LAPLACE'S, LAME'S, AND BESSEL'S FUNCTIONS. Crown 8vo. 10s. 6d.

Wilson (J. M.)—ELEMENTARY GEOMETRY. Books I. II. III. Containing the Subjects of Euclid's first Four Books. New Edition, following the Syllabus of the Geometrical Association. By J. M. WILSON, M.A., late Fellow of St. John's College, Cambridge, and Mathematical Master of Rugby School. Extra fcap. 8vo. 3s. 6d.

SOLID GEOMETRY AND CONIC SECTIONS. With Appendices on Transversals and Harmonic Division. For the use of Schools. By J. M. WILSON, M.A. Second Edition. Extra fcap. 8vo. 3s. 6d.

Wilson (W. P.)—A TREATISE ON DYNAMICS. By W. P. WILSON, M.A., Fellow of St. John's College, Cambridge, and Professor of Mathematics in Queen's College, Belfast. 8vo. 9s. 6d.

"*This treatise supplies a great educational need.*"—EDUCATIONAL TIMES.

Wolstenholme.—A BOOK OF MATHEMATICAL PROBLEMS, on Subjects included in the Cambridge Course. By JOSEPH WOLSTENHOLME, Fellow of Christ's College, sometime Fellow of St. John's College, and lately Lecturer in Mathematics at Christ's College. Crown 8vo. cloth. 8s. 6d.

"*Judicious, symmetrical, and well arranged.*"—GUARDIAN.

SCIENCE.

ELEMENTARY CLASS-BOOKS.

ASTRONOMY, by the Astronomer Royal.
POPULAR ASTRONOMY. With Illustrations. By SIR G. B. AIRY, K.C.B., Astronomer Royal. New Edition. 18mo. cloth. 4s. 6d.

Six lectures, intended " to explain to intelligent persons the principles on which the instruments of an Observatory are constructed, and the principles on which the observations made with these instruments are treated for deduction of the distances and weights of the bodies of the Solar System."

ASTRONOMY.
ELEMENTARY LESSONS IN ASTRONOMY. With Coloured Diagram of the Spectra of the Sun, Stars, and Nebulæ, and numerous Illustrations. By J. NORMAN LOCKYER, F.R.S. New Edition. 18mo. 5s. 6d.

"*Full, clear, sound, and worthy of attention, not only as a popular exposition, but as a scientific 'Index.'*"—ATHENÆUM. "*The most fascinating of elementary books on the Sciences.*"—NONCONFORMIST.

Elementary Class-Books—*continued.*

QUESTIONS ON LOCKYER'S ELEMENTARY LESSONS IN ASTRONOMY. For the Use of Schools. By JOHN FORBES-ROBERTSON. 18mo. cloth limp. 1*s.* 6*d.*

PHYSIOLOGY.

LESSONS IN ELEMENTARY PHYSIOLOGY. With numerous Illustrations. By T. H. HUXLEY, F.R.S., Professor of Natural History in the Royal School of Mines. New Edition. 18mo. cloth. 4*s.* 6*d.*

"*Pure gold throughout.*"—GUARDIAN. "*Unquestionably the clearest and most complete elementary treatise on this subject that we possess in any language.*"—WESTMINSTER REVIEW.

QUESTIONS ON HUXLEY'S PHYSIOLOGY FOR SCHOOLS. By T. ALCOCK, M.D. 18mo. 1*s.* 6*d.*

BOTANY.

LESSONS IN ELEMENTARY BOTANY. By D. OLIVER, F.R.S., F.L.S., Professor of Botany in University College, London. With nearly Two Hundred Illustrations. New Edition. 18mo. cloth. 4*s.* 6*d.*

CHEMISTRY.

LESSONS IN ELEMENTARY CHEMISTRY, INORGANIC AND ORGANIC. By HENRY E. ROSCOE, F.R.S., Professor of Chemistry in Owens College, Manchester. With numerous Illustrations and Chromo-Litho of the Solar Spectrum, and of the Alkalies and Alkaline Earths. New Edition. 18mo. cloth. 4*s.* 6*d.*

"*As a standard general text-book it deserves to take a leading place.*"—SPECTATOR. "*We unhesitatingly pronounce it the best of all our elementary treatises on Chemistry.*"—MEDICAL TIMES.

A SERIES OF CHEMICAL PROBLEMS, prepared [with Special Reference to the above, by T. E. THORPE, Ph.D., Professor of Chemistry in the Yorkshire College of Science, Leeds. Adapted for the preparation of Students for the Government, Science, and Society of Arts Examinations. With a Preface by Professor ROSCOE. 18mo. 1*s.* Key. 1*s.*

POLITICAL ECONOMY.

POLITICAL ECONOMY FOR BEGINNERS. By MILLICENT G. FAWCETT. New Edition. 18mo. 2*s.* 6*d.*

"*Clear, compact, and comprehensive.*"—DAILY NEWS. "*The relations of capital and labour have never been more simply or more clearly expounded.*"—CONTEMPORARY REVIEW.

LOGIC.

ELEMENTARY LESSONS IN LOGIC; Deductive and Inductive, with copious Questions and Examples, and a Vocabulary of

SCIENCE.

Elementary Class-Books—*continued.*
Logical Terms. By W. STANLEY JEVONS, M.A., Professor of Logic in University College, London. New Edition. 18mo. 3s. 6d.
"*Nothing can be better for a school-book.*"—GUARDIAN.
"*A manual alike simple, interesting, and scientific.*"—ATHENÆUM.

PHYSICS.
LESSONS IN ELEMENTARY PHYSICS. By BALFOUR STEWART, F.R.S., Professor of Natural Philosophy in Owens College, Manchester. With numerous Illustrations and Chromoliths of the Spectra of the Sun, Stars, and Nebulæ. New Edition. 18mo. 4s. 6d.
"*The beau-ideal of a scientific text-book, clear, accurate, and thorough.*"—EDUCATIONAL TIMES.

PRACTICAL CHEMISTRY.
THE OWENS COLLEGE JUNIOR COURSE OF PRACTICAL CHEMISTRY. By FRANCIS JONES, Chemical Master in the Grammar School, Manchester. With Preface by Professor ROSCOE, and Illustrations. New Edition. 18mo. 2s. 6d.

ANATOMY.
LESSONS IN ELEMENTARY ANATOMY. By ST. GEORGE MIVART, F.R.S., Lecturer in Comparative Anatomy at St. Mary's Hospital. With upwards of 400 Illustrations. 18mo. 6s. 6d.
"*It may be questioned whether any other work on Anatomy contains in like compass so proportionately great a mass of information.*"—LANCET.
"*The work is excellent, and should be in the hands of every student of human anatomy.*"—MEDICAL TIMES.

STEAM.—AN ELEMENTARY TREATISE. By JOHN PERRY, Bachelor of Engineering, Whitworth Scholar, etc., late Lecturer in Physics at Clifton College. With numerous Woodcuts and Numerical Examples and Exercises. 18mo. 4s. 6d.
"*The young engineer and those seeking for a comprehensive knowledge of the use, power, and economy of steam, could not have a more useful work, as it is very intelligible, well arranged, and practical throughout.*"—IRONMONGER.

PHYSICAL GEOGRAPHY.
ELEMENTARY LESSONS IN PHYSICAL GEOGRAPHY. By A. GEIKIE, F.R.S., Murchison Professor of Geology, etc., Edinburgh. With numerous Illustrations. 18mo. [*Shortly.*

MANUALS FOR STUDENTS.
Flower (W. H.)—AN INTRODUCTION TO THE OSTEOLOGY OF THE MAMMALIA. Being the substance of the Course of Lectures delivered at the Royal College of Surgeons of England in 1870. By W. H. FLOWER, F.R.S., F.R.C.S., Hunterian Professor of Comparative Anatomy and Physiology. With numerous Illustrations. Second Edition enlarged. Crown 8vo. 10s. 6d.

Foster & Balfour.—THE ELEMENTS OF EMBRYOLOGY. By MICHAEL FOSTER, M.D., F.R.S. & F. M. BALFOUR, M.A. Part I. crown 8vo. 7s. 6d.

Foster & Langley.—A COURSE OF ELEMENTARY PRACTICAL PHYSIOLOGY. By MICHAEL FOSTER, M.D., F.R.S., and J. N. LANGLEY, B.A. Crown 8vo. 6s.

Hooker (Dr.)—THE STUDENT'S FLORA OF THE BRITISH ISLANDS. By J. D. HOOKER, C.B., F.R.S., M.D., D.C.L., President of the Royal Society. Globe 8vo. 10s. 6d.

"*Cannot fail to perfectly fulfil the purpose for which it is intended.*"—LAND AND WATER. "*Certainly the fullest and most accurate manual of the kind that has yet appeared.*"—PALL MALL GAZETTE.

Huxley & Martin.—A COURSE OF PRACTICAL INSTRUCTION IN ELEMENTARY BIOLOGY. By Professor HUXLEY, F.R.S., assisted by H. N. MARTIN, M.B., D.Sc. Second Edition, revised. Crown 8vo. 6s.

"*It is impossible for an intelligent youth, with this book in his hand, placing himself before any one of the organisms described, and carefully following the directions given, to fail to verify each point to which his attention is directed.*"—ATHENÆUM.

Oliver (Professor).—FIRST BOOK OF INDIAN BOTANY. By DANIEL OLIVER, F.R.S., F.L.S., Keeper of the Herbarium and Library of the Royal Gardens, Kew, and Professor of Botany in University College, London. With numerous Illustrations. Extra fcap. 8vo. 6s. 6d.

"*It contains a well-digested summary of all essential knowledge pertaining to Indian botany, wrought out in accordance with the best principles of scientific arrangement.*"—ALLEN'S INDIAN MAIL.

Other volumes of these Manuals will follow.

NATURE SERIES.

THE SPECTROSCOPE AND ITS APPLICATIONS. By J. NORMAN LOCKYER, F.R.S. With Coloured Plate and numerous Illustrations. Second Edition. Crown 8vo. 3s. 6d.

THE ORIGIN AND METAMORPHOSES OF INSECTS. By SIR JOHN LUBBOCK, M.P., F.R.S., D.C.L. With numerous Illustrations. Second Edition. Crown 8vo. 3s. 6d.

"*We can most cordially recommend it to young naturalists.*"—ATHENÆUM.

THE TRANSIT OF VENUS. By G. FORBES, M.A., Professor of Natural Philosophy in the Andersonian University, Glasgow. Illustrated. Crown 8vo. 3s. 6d.

THE COMMON FROG. By ST. GEORGE MIVART, F.R.S., Lecturer in Comparative Anatomy at St. Mary's Hospital. With numerous Illustrations. Crown 8vo. 3s. 6d.

Nature Series—*continued.*

POLARISATION OF LIGHT. By W. SPOTTISWOODE, F.R.S. With many Illustrations. Second Edition. Crown 8vo. 3s. 6d.

ON BRITISH WILD FLOWERS CONSIDERED IN RELATION TO INSECTS. By SIR JOHN LUBBOCK, M.P., F.R.S. With numerous Illustrations. Second Edition. Crown 8vo. 4s. 6d.

Other volumes to follow.

Ball (R. S., A.M.)—EXPERIMENTAL MECHANICS. A Course of Lectures delivered at the Royal College of Science for Ireland. By R. S. BALL, A.M., Professor of Applied Mathematics and Mechanics in the Royal College of Science for Ireland. Royal 8vo. 16s.

Blanford.—THE RUDIMENTS OF PHYSICAL GEOGRAPHY FOR THE USE OF INDIAN SCHOOLS; with a Glossary of Technical Terms employed. By H. F. BLANFORD, F.R.S. Fifth edition, with Illustrations. Globe 8vo. 2s. 6d.

Gordon.—AN ELEMENTARY BOOK ON HEAT. By J. E. H. GORDON, B.A., Gonville and Caius College, Cambridge. Crown 8vo. 2s.

Reuleaux.—THE KINEMATICS OF MACHINERY. Outlines of a Theory of Machines. By Professor F. REULEAUX, Translated and Edited by Professor A. B. KENNEDY, C.E. With 450 Illustrations. Medium 8vo. 21s.

Roscoe and Schorlemmer.—CHEMISTRY. A Complete Treatise on. By Professor H. E. ROSCOE, F.R.S., and Professor C. SCHORLEMMER, F.R.S. With numerous Illustrations. Medium 8vo. [*Nearly ready.*

SCIENCE PRIMERS FOR ELEMENTARY SCHOOLS.

Under the joint Editorship of Professors HUXLEY, ROSCOE, and BALFOUR STEWART.

"*These Primers are extremely simple and attractive, and thoroughly answer their purpose of just leading the young beginner up to the threshold of the long avenues in the Palace of Nature which these titles suggest.*"—GUARDIAN. "*They are wonderfully clear and lucid in their instruction, simple in style, and admirable in plan.*"—EDUCATIONAL TIMES.

PRIMER OF CHEMISTRY. By H. E. ROSCOE, Professor of Chemistry in Owens College, Manchester. With numerous Illustrations. 18mo. 1s. New Edition. With Questions.

"*A very model of perspicacity and accuracy.*"—CHEMIST AND DRUGGIST.

PRIMER OF PHYSICS. By BALFOUR STEWART, Professor of Natural Philosophy in Owens College, Manchester. With numerous Illustrations. 18mo. 1s. New Edition. With Questions.

Science Primers—*continued.*

PRIMER OF PHYSICAL GEOGRAPHY. By ARCHIBALD GEIKIE, F.R.S., Murchison-Professor of Geology and Mineralogy at Edinburgh. With numerous Illustrations. New Edition. 18mo. 1s.

"*Everyone of his lessons is marked by simplicity, clearness, and correctness.*"—ATHENÆUM.

PRIMER OF GEOLOGY. By PROFESSOR GEIKIE, F.R.S. With numerous Illustrations. New Edition. 18mo. cloth. 1s.

"*It is hardly possible for the dullest child to misunderstand the meaning of a classification of stones after Professor Geikie's explanation.*"—SCHOOL BOARD CHRONICLE.

PRIMER OF PHYSIOLOGY. By MICHAEL FOSTER, M.D., F.R.S. With numerous Illustrations. New Edition. 18mo. 1s.

"*The book seems to us to leave nothing to be desired as an elementary text-book.*"—ACADEMY.

PRIMER OF ASTRONOMY. By J. NORMAN LOCKYER, F.R.S. With numerous Illustrations. New Edition. 18mo. 1s.

"*This is altogether one of the most likely attempts we have ever seen to bring astronomy down to the capacity of the young child.*"—SCHOOL BOARD CHRONICLE.

PRIMER OF BOTANY. By J. D. HOOKER, C.B., F.R.S., President of the Royal Society. With numerous Illustrations. New Edition. 18mo. 1s.

"*To teachers the Primer will be of inestimable value, and not only because of the simplicity of the language and the clearness with which the subject matter is treated, but also on account of its coming from the highest authority and so furnishing positive information as to the most suitable methods of teaching the science of botany.*"—NATURE.

PRIMER OF LOGIC. By PROFESSOR STANLEY JEVONS, F.R.S. 18mo. 1s.

In preparation:—

INTRODUCTORY. By PROFESSOR HUXLEY. &c. &c.

SCIENCE LECTURES AT SOUTH KENSINGTON.

With Illustrations. Crown 8vo. 6d. each.

SOUND AND MUSIC. By Dr. W. H. STONE.
PHOTOGRAPHY. By Captain ABNEY, R.E.
KINEMATIC MODELS. By Professor KENNEDY, C.E.
OUTLINES OF FIELD GEOLOGY. By Professor GEIKIE, F.R.S.
ABSORPTION OF LIGHT, AND FLUORESCENCE. By Professor STOKES, F.R.S.

Others to follow.

MANCHESTER SCIENCE LECTURES FOR THE PEOPLE.

Eight Series, 1876-7. Crown 8vo. Illustrated. 6d. each.

WHAT THE EARTH IS COMPOSED OF. By Professor ROSCOE, F.R.S.
THE SUCCESSION OF LIFE ON THE EARTH. By Professor WILLIAMSON, F.R.S

Others to follow.

MISCELLANEOUS.

Abbott.—A SHAKESPEARIAN GRAMMAR. An Attempt to illustrate some of the Differences between Elizabethan and Modern English. By the Rev. E. A. ABBOTT, M.A., Head Master of the City of London School. New Edition. Extra fcap. 8vo. 6*s*.

"*Valuable not only as an aid to the critical study of Shakespeare, but as tending to familiarize the reader with Elizabethan English in general.*"—ATHENÆUM.

Barker.—FIRST LESSONS IN THE PRINCIPLES OF COOKING. By LADY BARKER. New Edition. 18mo. 1*s*.

"*An unpretending but invaluable little work .... The plan is admirable in its completeness and simplicity; it is hardly possible that anyone who can read at all can fail to understand the practical lessons on bread and beef, fish and vegetables.*"—SPECTATOR.

Berners.—FIRST LESSONS ON HEALTH. By J. BERNERS. Seventh Edition. 18mo. 1*s*.

Breymann.—Works by HERMANN BREYMANN, Ph.D., Professor of Philology in the University of Munich.

A FRENCH GRAMMAR BASED ON PHILOLOGICAL PRINCIPLES. Second Edition. Extra fcap. 8vo. 4*s*. 6*d*.

"*A good, sound, valuable philological grammar.*"—SCHOOL BOARD CHRONICLE.

FIRST FRENCH EXERCISE BOOK. Extra fcap. 8vo. 4*s*. 6*d*.

SECOND FRENCH EXERCISE BOOK. Extra fcap. 8vo. 2*s*. 6*d*.

Calderwood.—HANDBOOK OF MORAL PHILOSOPHY. By the Rev. HENRY CALDERWOOD, LL.D., Professor of Moral Philosophy, University of Edinburgh. Fourth Edition. Crown 8vo. 6*s*.

"*A compact and useful work .... will be an assistance to many students outside the author's own University.*"—GUARDIAN.

Delamotte.—A BEGINNER'S DRAWING BOOK. By P. H. DELAMOTTE, F.S.A. Progressively arranged. New Edition, improved. Crown 8vo. 3*s*. 6*d*.

"*A concise, simple, and thoroughly practical work.*"—GUARDIAN.

Fawcett.—TALES IN POLITICAL ECONOMY. By MILLICENT GARRETT FAWCETT. Globe 8vo. 3*s*.

"*The idea is a good one, and it is quite wonderful what a mass of economic teaching the author manages to compress into a small space.*"—ATHENÆUM.

Fearon.—SCHOOL INSPECTION, By D. R. FEARON, M.A., Assistant Commissioner of Endowed Schools. Second Edition. Crown 8vo. 2*s*. 6*d*.

"*The work is admirably adapted to serve the purpose for which it has been written. It is calculated to be eminently useful, and to have a powerful influence for good on our elementary education.*"—ATHENÆUM.

Fleay.—A SHAKESPEARE MANUAL. By F. G. FLEAY, M.A., Head Master of Skipton Grammar School. Extra fcap. 8vo. 4s. 6d.

"*A valuable contribution to the study of Shakespeare.*"—SATURDAY REVIEW.

Goldsmith.—THE TRAVELLER, or a Prospect of Society; and THE DESERTED VILLAGE. By OLIVER GOLDSMITH. With Notes Philological and Explanatory, by J. W. HALES, M.A. Crown 8vo. 6d.

Hales.—LONGER ENGLISH POEMS, with Notes, Philological and Explanatory, and an Introduction on the Teaching of English. Chiefly for use in Schools. Edited by J. W. HALES, M.A., Lecturer in English Literature and Classical Composition at King's College School, London, &c., &c. Third Edition. Extra fcap. 8vo. 4s. 6d.

"*The notes are very full and good, and the book, edited by one of our most cultivated English scholars, is probably the best volume of selections ever made for the use of English schools.*"—PROFESSOR MORLEY'S *First Sketch of English Literature.*

Hole.—A GENEALOGICAL STEMMA OF THE KINGS OF ENGLAND AND FRANCE. By the Rev. C. HOLE. On Sheet. 1s.

Jephson.—SHAKESPEARE'S "TEMPEST." With Glossarial and Explanatory Notes. By the Rev. J. M. JEPHSON. Second Edition. 18mo. 1s.

Literature Primers.—Edited by JOHN RICHARD GREEN. Author of "A Short History of the English People."

ENGLISH GRAMMAR. By the Rev. R. MORRIS, LL.D., President of the Philological Society. New Edition. 18mo. cloth. 1s.

"*A work quite precious in its way. . . . An excellent English Grammar for the lowest form.*"—EDUCATIONAL TIMES.

THE CHILDREN'S TREASURY OF LYRICAL POETRY. Selected and arranged with Notes by FRANCIS TURNER PALGRAVE. In Two Parts. 18mo. 1s. each.

ENGLISH LITERATURE. By the Rev. STOPFORD BROOKE, M.A. New Edition. 18mo. 1s.

"*Unquestionably the best short sketch of English literature that has appeared.*"—ATHENÆUM.

PHILOLOGY. By J. PEILE, M.A. 18mo. 1s.

In preparation :—
    LATIN LITERATURE. By the Rev. Dr. FARRAR, F.R.S.
    GREEK LITERATURE. By PROFESSOR JEBB, M.A.
    SHAKSPERE. By PROFESSOR DOWDEN.
    BIBLE PRIMER. By G. GROVE, D.C.L.
    CHAUCER. By F. J. FURNIVALL, M.A.

MISCELLANEOUS.

Martin.—THE POET'S HOUR: Poetry Selected and Arranged for Children. By FRANCES MARTIN. Second Edition. 18mo. 2s. 6d.
SPRING-TIME WITH THE POETS: Poetry selected by FRANCES MARTIN. Second Edition. 18mo. 3s. 6d.

Masson (Gustave).—A COMPENDIOUS DICTIONARY OF THE FRENCH LANGUAGE (French-English and English-French). Followed by a List of the Principal Diverging Derivations, and preceded by Chronological and Historical Tables. By GUSTAVE MASSON, Assistant-Master and Librarian, Harrow School. Third Edition. Square half-bound, 6s.

"*A book which any student, whatever may be the degree of his advancement in the language, would do well to have on the table close at hand while he is reading.*"—SATURDAY REVIEW.

Morris.—Works by the Rev. R. MORRIS, LL.D., Lecturer on English Language and Literature in King's College School.
HISTORICAL OUTLINES OF ENGLISH ACCIDENCE, comprising Chapters on the History and Development of the Language, and on Word-formation. Third Edition. Extra fcap. 8vo. 6s.

"*It marks an era in the study of the English tongue.*"—SATURDAY REVIEW. "*A genuine and sound book.*"—ATHENÆUM.
ELEMENTARY LESSONS IN HISTORICAL ENGLISH GRAMMAR, Containing Accidence and Word-formation. Second Edition. 18mo. 2s. 6d.
PRIMER OF ENGLISH GRAMMAR. 18mo. 1s.

Oliphant.—THE SOURCES OF STANDARD ENGLISH. By J. KINGTON OLIPHANT. Extra fcap. 8vo. 6s.

"*Comes nearer to a history of the English language than anything that we have seen since such a history could be written without confusion and contradictions.*"—SATURDAY REVIEW.

Palgrave.—THE CHILDREN'S TREASURY OF LYRICAL POETRY. Selected and Arranged with Notes by FRANCIS TURNER PALGRAVE. 18mo. 2s. 6d. Also in Two Parts. 18mo. 1s. each.

"*While indeed a treasure for intelligent children, it is also a work which many older folk will be glad to have.*"—SATURDAY REVIEW.

Pylodet.—NEW GUIDE TO GERMAN CONVERSATION: containing an Alphabetical List of nearly 800 Familiar Words followed by Exercises, Vocabulary of Words in frequent use, Familiar Phrases and Dialogues; a Sketch of German Literature, Idiomatic Expressions, &c. By L. PYLODET. 18mo. cloth limp. 2s. 6d.
A SYNOPSIS OF GERMAN GRAMMAR. By L. PYLODET. 18mo. 6d.

Reading Books.—Adapted to the English and Scotch Codes for 1875. Bound in Cloth.

PRIMER. 18mo. (48 pp.) 2*d*.
BOOK I. for Standard I. 18mo. (96 pp.) 4*d*.
  ,, II.   ,,   II. 18mo. (144 pp.) 5*d*.
  ,, III.   ,,   III. 18mo. (160 pp.) 6*d*.
  ,, IV.   ,,   IV. 18mo. (176 pp.) 8*d*.
  ,, V.   ,,   V. 18mo. (380 pp.) 1*s*.
  ,, VI.   ,,   VI. Crown 8vo. (430 pp.) 2*s*.

Book VI. is fitted for higher Classes, and as an Introduction to English Literature.

"*They are far above any others that have appeared both in form and substance.... The editor of the present series has rightly seen that reading books must 'aim chiefly at giving to the pupils the power of accurate, and, if possible, apt and skilful expression; at cultivating in them a good literary taste, and at arousing a desire of further reading.' This is done by taking care to select the extracts from true English classics, going up in Standard VI. course to Chaucer, Hooker, and Bacon, as well as Wordsworth, Macaulay, and Froude..... This is quite on the right track, and indicates justly the ideal which we ought to set before us.*"—GUARDIAN.

Skeat.—SHAKESPEARE'S PLUTARCH. Being a Selection from the Lives in North's Plutarch which illustrate Shakespeare's Plays. Edited with Introduction, Notes, Index of Names, and Glossarial Index. By the Rev. W. W. SKEAT, M.A. Crown 8vo. 6*s*.

Sonnenschein and Meiklejohn.—THE ENGLISH METHOD OF TEACHING TO READ. By A. SONNENSCHEIN and J. M. D. MEIKLEJOHN, M.A. Fcap. 8vo.

COMPRISING :

THE NURSERY BOOK, containing all the Two-Letter Words in the Language. 1*d*. (Also in Large Type on Sheets for School Walls. 5*s*.)

THE FIRST COURSE, consisting of Short Vowels with Single Consonants. 3*d*.

THE SECOND COURSE, with Combinations and Bridges, consisting of Short Vowels with Double Consonants. 4*d*.

THE THIRD AND FOURTH COURSES, consisting of Long Vowels, and all the Double Vowels in the Language. 6*d*.

"*These are admirable books, because they are constructed on a principle, and that the simplest principle on which it is possible to learn to read English.*"—SPECTATOR.

Taylor.—WORDS AND PLACES; or, Etymological Illustrations of History, Ethnology, and Geography. By the Rev. ISAAC TAYLOR, M.A. Third and cheaper Edition, revised and compressed. With Maps. Globe 8vo. 6*s*.

MISCELLANEOUS.

Tegetmeier.—THE SCHOLAR'S HANDBOOK OF HOUSE-HOLD MANAGEMENT AND COOKERY, SUITABLE FOR ELEMENTARY SCHOOLS. With an Appendix of Recipes used by the Teachers of the National School of Cookery. By W. B. TEGETMEIER. Compiled at the request of the School Board for London. 18mo. 1s.

Thring.—Works by EDWARD THRING, M.A., Head Master of Uppingham.
THE ELEMENTS OF GRAMMAR TAUGHT IN ENGLISH. With Questions. Fourth Edition. 18mo. 2s.
THE CHILD'S GRAMMAR. Being the Substance of "The Elements of Grammar taught in English," adapted for the Use of Junior Classes. A New Edition. 18mo. 1s.
SCHOOL SONGS. A Collection of Songs for Schools. With the Music arranged for four Voices. Edited by the Rev. E. THRING and H. RICCIUS. Folio. 7s. 6d.

Trench (Archbishop).—Works by R. C. TRENCH, D.D., Archbishop of Dublin.
HOUSEHOLD BOOK OF ENGLISH POETRY. Selected and Arranged, with Notes. Extra fcap. 8vo. 5s. 6d. Second Edition.
"*The Archbishop has conferred in this delightful volume an important gift on the whole English-speaking population of the world.*"—PALL MALL GAZETTE.
ON THE STUDY OF WORDS. Lectures addressed (originally) to the Pupils at the Diocesan Training School, Winchester. Sixteenth Edition, revised. Fcap. 8vo. 5s.
ENGLISH, PAST AND PRESENT. Ninth Edition, revised and improved. Fcap. 8vo. 5s.
A SELECT GLOSSARY OF ENGLISH WORDS, used formerly in Senses Different from their Present. Fourth Edition, enlarged. Fcap. 8vo. 4s. 6d.

Vaughan (C. M.)— A SHILLING BOOK OF WORDS FROM THE POETS. By C. M. VAUGHAN. 18mo. cloth.

Whitney.—Works by WILLIAM D. WHITNEY, Professor of Sanskrit and Instructor in Modern Languages in Yale College; first President of the American Philological Association, and hon. member of the Royal Asiatic Society of Great Britain and Ireland; and Correspondent of the Berlin Academy of Sciences.
A COMPENDIOUS GERMAN GRAMMAR. Crown 8vo. 6s.
A GERMAN READER IN PROSE AND VERSE, with Notes and Vocabulary. Crown 8vo. 7s. 6d.

Yonge (Charlotte M.)—THE ABRIDGED BOOK OF GOLDEN DEEDS. A Reading Book for Schools and General Readers. By the Author of "The Heir of Redclyffe." 18mo. cloth. 1s.

HISTORY.

Freeman (Edward A.)—OLD-ENGLISH HISTORY. By EDWARD A. FREEMAN, D.C.L., late Fellow of Trinity College, Oxford. With Five Coloured Maps. Fourth Edition. Extra fcap. 8vo. half-bound. 6s.

"*The book indeed is full of instruction and interest to students of all ages, and he must be a well-informed man indeed who will not rise from its perusal with clearer and more accurate ideas of a too much neglected portion of English History.*"—SPECTATOR.

Green.—A SHORT HISTORY OF THE ENGLISH PEOPLE. By JOHN RICHARD GREEN. With Coloured Maps, Genealogical Tables, and Chronological Annals. Crown 8vo. 8s. 6d. Forty-third Thousand.

"*Stands alone as the one general history of the country, for the sake of which all others, if young and old are wise, will be speedily and surely set aside.*"—ACADEMY.

Historical Course for Schools.—Edited by EDWARD A. FREEMAN, D.C.L., late Fellow of Trinity College, Oxford.

I. GENERAL SKETCH OF EUROPEAN HISTORY. By EDWARD A. FREEMAN, D.C.L. Fifth Edition, revised and enlarged, with Chronological Table, Maps, and Index. 18mo. cloth. 3s. 6d.

"*It supplies the great want of a good foundation for historical teaching. The scheme is an excellent one, and this instalment has been executed in a way that promises much for the volumes that are yet to appear.*"—EDUCATIONAL TIMES.

II. HISTORY OF ENGLAND. By EDITH THOMPSON. Fifth Edition. 18mo. 2s. 6d.

"*Freedom from prejudice, simplicity of style, and accuracy of statement, are the characteristics of this little volume. It is a trustworthy text-book and likely to be generally serviceable in schools.*"—PALL MALL GAZETTE. "*Upon the whole, this manual is the best sketch of English history for the use of young people we have yet met with.*"—ATHENÆUM.

III. HISTORY OF SCOTLAND. By MARGARET MACARTHUR. Second Edition. 18mo. 2s.

"*An excellent summary, unimpeachable as to facts, and putting them in the clearest and most impartial light attainable.*"—GUARDIAN. "*Miss Macarthur has performed her task with admirable care, clearness, and fulness, and we have now for the first time a really good School History of Scotland.*"—EDUCATIONAL TIMES.

IV. HISTORY OF ITALY. By the Rev. W. HUNT, M.A. 18mo. 3s.

"*It possesses the same solid merit as its predecessors .... the same scrupulous care about fidelity in details. ... It is distinguished, too, by information on art, architecture, and social politics, in which the writer's grasp is seen by the firmness and clearness of his touch.*"—EDUCATIONAL TIMES.

HISTORY.

Historical Course for Schools—*continued.*

V. HISTORY OF GERMANY. By J. SIME, M.A. 18mo. 3s.

"*A remarkably clear and impressive History of Germany. Its great events are wisely kept as central figures, and the smaller events are carefully kept, not only subordinate and subservient, but most skilfully woven into the texture of the historical tapestry presented to the eye.*"—STANDARD.

VI. HISTORY OF AMERICA. By JOHN A. DOYLE. With Maps. 18mo. 4s. 6d.

"*Mr. Doyle has performed his task with admirable care, fulness, and clearness, and for the first time we have for schools an accurate and interesting history of America, from the earliest to the present time.*"—STANDARD.

The following will shortly be issued :—
    FRANCE. By CHARLOTTE M. YONGE.
    GREECE. By J. ANNAN BRYCE, B.A.

History Primers.—Edited by JOHN RICHARD GREEN. Author of "A Short History of the English People."

ROME. By the Rev. M. Creighton, M.A., Fellow and Tutor of Merton College, Oxford. With Eleven Maps. New Edition. 18mo. 1s.

"*The Author has been curiously successful in telling in an intelligent way the story of Rome from first to last.*"—SCHOOL BOARD CHRONICLE.

GREECE. By C. A. Fyffe, M.A., Fellow and late Tutor of University College, Oxford. With Five Maps. New Edition. 18mo. 1s.

"*We give our unqualified praise to this little manual.*"—SCHOOLMASTER.

EUROPEAN HISTORY. By E. A. FREEMAN, D.C.L., LL.D. With Maps. New Edition. 18mo. 1s.

"*A marvel of clearness.*"—ACADEMY.

CLASSICAL ANTIQUITIES. I, OLD GREEK LIFE. By the Rev. J. P. MAHAFFY, M.A. Illustrated. 18mo. 1s.

CLASSICAL GEOGRAPHY. By H. F. TOZER, M.A. 18mo. 1s.

GEOGRAPHY. By GEORGE GROVE, D.C.L. With Maps. 18mo. 1s.

In preparation :—
    ENGLAND. By J. R. GREEN, M.A.
    FRANCE. By CHARLOTTE M. YONGE.

Michelet.—A SUMMARY OF MODERN HISTORY. Translated from the French of M. Michelet, and continued to the Present Time, by M. C. M. Simpson. Globe 8vo. 4s. 6d.

"*We are glad to see one of the ablest and most useful summaries of European history put into the hands of English readers. The translation is excellent.*"—STANDARD.

Otté.—SCANDINAVIAN HISTORY. By E. C. OTTÉ. With Maps. Globe 8vo. 6s.

"*A readable, well-arranged, complete, and accurate volume.*"—LITERARY REVIEW.

Pauli.—PICTURES OF OLD ENGLAND. By Dr. R. PAULI. Translated with the Sanction of the Author by E. C. OTTÉ. Cheaper Edition. Crown 8vo. 6s.

Yonge (Charlotte M.)—A PARALLEL HISTORY OF FRANCE AND ENGLAND: consisting of Outlines and Dates. By CHARLOTTE M. YONGE, Author of "The Heir of Redclyffe," "Cameos of English History," &c. &c. Oblong 4to. 3s. 6d.

"*We can imagine few more really advantageous courses of historical study for a young mind than going carefully and steadily through Miss Yonge's excellent little book.*"—EDUCATIONAL TIMES.

CAMEOS FROM ENGLISH HISTORY. From Rollo to Edward II. By the Author of "The Heir of Redclyffe." Extra fcap. 8vo. Third Edition, enlarged. 5s.

"*Instead of dry details, we have living pictures, faithful, vivid, and striking.*"—NONCONFORMIST.

A SECOND SERIES OF CAMEOS FROM ENGLISH HISTORY. THE WARS IN FRANCE. Third Edition. Extra fcap. 8vo. 5s.

"*Though mainly intended for young readers, they will, if we mistake not, be found very acceptable to those of more mature years, and the life and reality imparted to the dry bones of history cannot fail to be attractive to readers of every age.*"—JOHN BULL.

A THIRD SERIES OF CAMEOS FROM ENGLISH HISTORY. THE WARS OF THE ROSES. Extra fcap. 8vo. 5s.

EUROPEAN HISTORY. Narrated in a Series of Historical Selections from the Best Authorities. Edited and arranged by E. M. SEWELL and C. M. YONGE. First Series, 1003—1154. Third Edition. Crown 8vo. 6s. Second Series, 1088—1228. Crown 8vo. 6s. Third Edition.

"*We know of scarcely anything which is so likely to raise to a higher level the average standard of English education.*"—GUARDIAN.

DIVINITY.

∗∗* For other Works by these Authors, see THEOLOGICAL CATALOGUE.

Abbott (Rev. E. A.)—BIBLE LESSONS. By the Rev. E. A. ABBOTT, M.A., Head Master of the City of London School. Second Edition. Crown 8vo. 4s. 6d.

"*Wise, suggestive, and really profound initiation into religious thought.*" —GUARDIAN. "*I think nobody could read them without being both the better for them himself, and being also able to see how this difficult duty of imparting a sound religious education may be effected.*"—BISHOP OF ST. DAVID'S AT ABERGWILLY.

Arnold.— A BIBLE-READING FOR SCHOOLS. THE GREAT PROPHECY OF ISRAEL'S RESTORATION (Isaiah, Chapters xl.—lxvi.). Arranged and Edited for Young Learners. By MATTHEW ARNOLD, D.C.L., formerly Professor of Poetry in the University of Oxford, and Fellow of Oriel. Fourth Edition. 18mo. cloth. 1s.

"*There can be no doubt that it will be found excellently calculated to further instruction in Biblical literature in any school into which it may be introduced; and we can safely say that whatever school uses the book, it will enable its pupils to understand Isaiah, a great advantage compared with other establishments which do not avail themselves of it.*"—TIMES.

Arnold.—ISAIAH XL.—LXVI. With the Shorter Prophecies allied to it. Arranged and Edited with Notes by MATTHEW ARNOLD. Crown 8vo. 5s.

Golden Treasury Psalter.—Students' Edition. Being an Edition of "The Psalms Chronologically Arranged, by Four Friends," with briefer Notes. 18mo. 3s. 6d.

Hardwick.—A HISTORY OF THE CHRISTIAN CHURCH. Middle Age. From Gregory the Great to the Excommunication of Luther. Edited by WILLIAM STUBBS, M.A., Regius Professor of Modern History in the University of Oxford. With Four Maps constructed for this work by A. KEITH JOHNSTON. Fourth Edition. Crown 8vo. 10s. 6d.

"*As a manual for the student of ecclesiastical history in the Middle Ages, we know no English work which can be compared to Mr. Hardwick's book.*"—GUARDIAN.

A HISTORY OF THE CHRISTIAN CHURCH DURING THE REFORMATION. By ARCHDEACON HARDWICK. Fourth Edition. Edited by Professor STUBBS. Crown 8vo. 10s. 6d.

Maclear.—Works by the Rev. G. F. MACLEAR, D.D., Head Master of King's College School.

A CLASS-BOOK OF OLD TESTAMENT HISTORY. Ninth Edition, with Four Maps. 18mo. cloth. 4s. 6d.

"*A careful and elaborate though brief compendium of all that modern research has done for the illustration of the Old Testament. We know of no work which contains so much important information in so small a compass.*"—BRITISH QUARTERLY REVIEW.

A CLASS-BOOK OF NEW TESTAMENT HISTORY, including the Connexion of the Old and New Testament. With Four Maps. Sixth Edition. 18mo. cloth. 5s. 6d.

"*A singularly clear and orderly arrangement of the Sacred Story. His work is solidly and completely done.*"—ATHENÆUM.

A SHILLING BOOK OF OLD TESTAMENT HISTORY, for National and Elementary Schools. With Map. 18mo. cloth. New Edition.

Maclear—*continued.*

A SHILLING BOOK OF NEW TESTAMENT HISTORY, for National and Elementary Schools. With Map. 18mo. cloth. New Edition.

These works have been carefully abridged from the author's larger manuals.

CLASS-BOOK OF THE CATECHISM OF THE CHURCH OF ENGLAND. New and Cheaper Edition. 18mo. cloth. 1s. 6d.

"*It is indeed the work of a scholar and divine, and as such, though extremely simple, it is also extremely instructive. There are few clergymen who would not find it useful in preparing candidates for Confirmation; and there are not a few who would find it useful to themselves as well.*"—LITERARY CHURCHMAN.

A FIRST CLASS-BOOK OF THE CATECHISM OF THE CHURCH OF ENGLAND, with Scripture Proofs, for Junior Classes and Schools. 18mo. 6d. New Edition.

A MANUAL OF INSTRUCTION FOR CONFIRMATION AND FIRST COMMUNION. With Prayers and Devotions. Royal 32mo. cloth extra, red edges. 2s.

"*It is earnest, orthodox, and affectionate in tone. The form of self-examination is particularly good.*"—JOHN BULL.

THE ORDER OF CONFIRMATION, WITH PRAYERS AND DEVOTIONS. 32mo. 6d.

FIRST COMMUNION, WITH PRAYERS AND DEVOTIONS FOR THE NEWLY CONFIRMED. 32mo. 6d.

Maurice.—THE LORD'S PRAYER, THE CREED, AND THE COMMANDMENTS. A Manual for Parents and Schoolmasters. To which is added the Order of the Scriptures. By the Rev. F. DENISON MAURICE, M.A. 18mo. cloth limp. 1s.

Procter.—A HISTORY OF THE BOOK OF COMMON PRAYER, with a Rationale of its Offices. By FRANCIS PROCTER, M.A. Twelfth Edition, revised and enlarged. Crown 8vo. 10s. 6d.

Procter and Maclear.—AN ELEMENTARY INTRODUCTION TO THE BOOK OF COMMON PRAYER. Re-arranged and supplemented by an Explanation of the Morning and Evening Prayer and the Litany. By the Rev. F. PROCTER and the Rev. Dr. MACLEAR. New Edition. 18mo. 2s. 6d.

Psalms of David Chronologically Arranged. By Four Friends. An Amended Version, with Historical Introduction and Explanatory Notes. Second and Cheaper Edition, with Additions and Corrections. Crown 8vo. 8s. 6d.

"*One of the most instructive and valuable books that has been published for many years.*"—SPECTATOR.

Ramsay.—THE CATECHISER'S MANUAL; or, the Church Catechism Illustrated and Explained, for the use of Clergymen, Schoolmasters, and Teachers. By the Rev. ARTHUR RAMSAY, M.A. Second Edition. 18mo. 1s. 6d.

Simpson.—AN EPITOME OF THE HISTORY OF THE CHRISTIAN CHURCH. By WILLIAM SIMPSON, M.A. Fifth Edition. Fcap. 8vo. 3s. 6d.

Swainson.—A HANDBOOK to BUTLER'S ANALOGY. By C. A. SWAINSON, D.D., Canon of Chichester. Crown 8vo. 1s. 6d.

Trench.—SYNONYMS OF THE NEW TESTAMENT. By R. CHENEVIX TRENCH, D.D., Archbishop of Dublin. Eighth Edition, revised. 8vo. 12s.

Westcott.—Works by BROOKE FOSS WESTCOTT, D.D., Canon of Peterborough.

A GENERAL SURVEY OF THE HISTORY OF THE CANON OF THE NEW TESTAMENT DURING THE FIRST FOUR CENTURIES. Fourth Edition. With Preface on "Supernatural Religion." Crown 8vo. 10s. 6d.

"*As a theological work it is at once perfectly fair and impartial, and imbued with a thoroughly religious spirit; and as a manual it exhibits, in a lucid form and in a narrow compass, the results of extensive research and accurate thought. We cordially recommend it.*"—SATURDAY REVIEW.

INTRODUCTION TO THE STUDY OF THE FOUR GOSPELS. Fifth Edition. Crown 8vo. 10s. 6d.

"*To learning and accuracy which commands respect and confidence, he unites what are not always to be found in union with these qualities, the no less valuable faculties of lucid arrangement and graceful and facile expression.*"—LONDON QUARTERLY REVIEW.

THE BIBLE IN THE CHURCH. A Popular Account of the Collection and Reception of the Holy Scriptures in the Christian Churches. New Edition. 18mo. cloth. 4s. 6d.

"*We would recommend every one who loves and studies the Bible to read and ponder this exquisite little book. Mr. Westcott's account of the 'Canon' is true history in its highest sense.*"—LITERARY CHURCHMAN.

THE GOSPEL OF THE RESURRECTION. Thoughts on its Relation to Reason and History. New Edition. Crown 8vo. 6s.

Wilson.—THE BIBLE STUDENT'S GUIDE to the more Correct Understanding of the English translation of the Old Testament, by reference to the Original Hebrew. By WILLIAM WILSON, D.D., Canon of Winchester, late Fellow of Queen's College, Oxford. Second Edition, carefully Revised. 4to. cloth. 25s.

"*For all earnest students of the Old Testament Scriptures it is a most valuable Manual. Its arrangement is so simple that those who possess only their mother-tongue, if they will take a little pains, may employ it with great profit.*"—NONCONFORMIST.

Yonge (Charlotte M.)—SCRIPTURE READINGS FOR SCHOOLS AND FAMILIES. By CHARLOTTE M. YONGE, Author of "The Heir of Redclyffe." FIRST SERIES. Genesis to Deuteronomy. Globe 8vo. 1s. 6d. With Comments. Second Edition. 3s. 6d.

SECOND SERIES. From JOSHUA to SOLOMON. Extra fcap. 8vo. 1s. 6d. With Comments, 3s. 6d.

THIRD SERIES. The KINGS and the PROPHETS. Extra fcap. 8vo. 1s. 6d. With Comments, 3s. 6d.

FOURTH SERIES. The GOSPEL TIMES. 1s. 6d. With Comments Extra fcap. 8vo. 3s. 6d.

Actual need has led the author to endeavour to prepare a reading book convenient for study with children, containing the very words of the Bible, with only a few expedient omissions, and arranged in Lessons of such length as by experience she has found to suit with children's ordinary power of accurate attentive interest. The verse form has been retained, because of its convenience for children reading in class, and as more resembling their Bibles; but the poetical portions have been given in their lines. When Psalms or portions from the Prophets illustrate or fall in with the narrative they are given in their chronological sequence. The Scripture portion, with a very few notes explanatory of mere words, is bound up apart, to be used by children, while the same is also supplied with a brief comment, the purpose of which is either to assist the teacher in explaining the lesson, or to be used by more advanced young people to whom it may not be possible to give access to the authorities whence it has been taken. Professor Huxley, at a meeting of the London School Board, particularly mentioned the selection made by Miss Yonge as an example of how selections might be made from the Bible for School Reading. See TIMES, *March* 30, 1871.

www.ingramcontent.com/pod-product-compliance
Lightning Source LLC
Chambersburg PA
CBHW021806230426
43669CB00008B/644